Mary Ball Washington

Mary Ball Washington

THE UNTOLD STORY OF
GEORGE WASHINGTON'S MOTHER

Craig Shirley

HARPER

An Imprint of HarperCollinsPublishers

HarperCollins books may be purchased for educational, business, or sales promotional use. For information, please email the Special Markets Department at SPsales@harpercollins.com.

FIRST EDITION

Designed by Leah Carlson-Stanisic

Frontispiece: Washington's Farewell to his Mother, 1789 *by Jean Leon Gerome Ferris © Virginia Museum of History & Culture*

Library of Congress Cataloging-in-Publication Data has been applied for.

ISBN 978-0-06-245651-9

19 20 21 22 23   LSC   10 9 8 7 6 5 4 3 2 1

Dedicated to some of the strong and smart and capable and resilient women I've come to know, love, and respect . . .

Barbara Shirley Eckert, Taylor Shirley Gillespie, Rebecca Shirley Sirhal, Diana Banister, Laura Ingraham, Gay Hart Gaines, Joanne Herring, Karin Andrews, Brittany Singer, Torrence Harman, Georgette Mosbacher, Mari Will, Callista Gingrich, Susan McShane, Soonalyn Jacob, Candy Bhappu, Carmen Bhappu, Ruby Jamshedi, Chris Kabanuk, Ellen Shirley, Amy Mauer, and Erin Mauer, but most especially for . . .

my best friend, my soul mate, my best editor, my love, my wife, Zorine Bhappu Shirley.

I was often there with George, his playmate, schoolmate,
and young man's companion. Of the mother I was ten
times more afraid than of my own parents. She awed me
in the midst of her kindness, for she was indeed truly
kind, . . . and even now, when time has whitened my locks,
and I am the grand-father of a second generation, I could
not behold that majestic woman without feelings it is
impossible to describe.

—LAWRENCE WASHINGTON ESQ., OF CHOTANK,
COUSIN OF GEORGE WASHINGTON

Contents

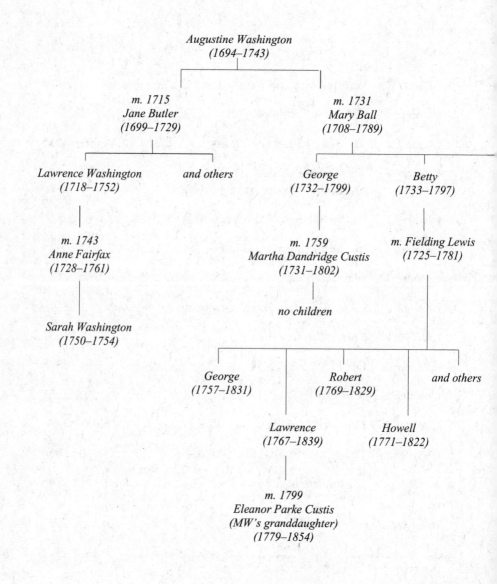

Augustine Washington
(1694–1743)

m. 1715
Jane Butler
(1699–1729)

m. 1731
Mary Ball
(1708–1789)

Lawrence Washington
(1718–1752)

and others

George
(1732–1799)

Betty
(1733–1797)

m. 1743
Anne Fairfax
(1728–1761)

m. 1759
Martha Dandridge Custis
(1731–1802)

m. Fielding Lewis
(1725–1781)

no children

Sarah Washington
(1750–1754)

George
(1757–1831)

Robert
(1769–1829)

and others

Lawrence
(1767–1839)

Howell
(1771–1822)

m. 1799
Eleanor Parke Custis
(MW's granddaughter)
(1779–1854)

The Washington Family

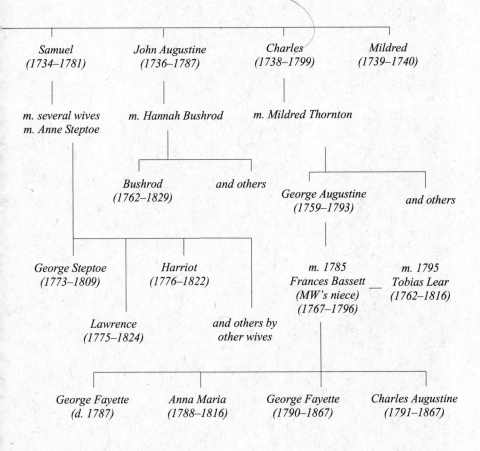

Samuel
(1734–1781)

John Augustine
(1736–1787)

Charles
(1738–1799)

Mildred
(1739–1740)

m. several wives
m. Anne Steptoe

m. Hannah Bushrod

m. Mildred Thornton

Bushrod
(1762–1829)

and others

George Augustine
(1759–1793)

and others

George Steptoe
(1773–1809)

Harriot
(1776–1822)

m. 1785
Frances Bassett
(MW's niece)
(1767–1796)

m. 1795
Tobias Lear
(1762–1816)

Lawrence
(1775–1824)

and others by
other wives

George Fayette
(d. 1787)

Anna Maria
(1788–1816)

George Fayette
(1790–1867)

Charles Augustine
(1791–1867)

News of Yorktown, 1781 by Jean Leon Gerome Ferris © Virginia Museum of History & Culture

Mary Ball Washington

Prologue

W hen General George Washington successfully led the American Revolution against the most powerful military force in the world, that of the British Empire, King George III reportedly said that if Washington laid down his sword after his stunning, startling, and world-altering victory, he would be regarded as the greatest man in the world.

That is precisely what Washington did.

At the conclusion of the Revolutionary War, he went before the Continental Congress in Annapolis, Maryland, on December 23, 1783, laid down his weapon, and made a brief speech. After celebrations and after commemorations, he left and went home to become, again, Planter Washington.

He was eager to return to his beloved wife, Martha, and his cherished Mount Vernon—which he hadn't been to in seven years, except briefly to entertain some French officers while on his way to Yorktown and a date with destiny against General Charles Cornwallis—to get back to the life of the gentleman farmer he so loved. In character, he gave the credit for the miracle of the American Revolution to others.

"Your excellency is retired like [another] Cincinnatus," wrote the Marquis de Chastellux at the close of the Revolutionary War.[1] Lucius Quinctius Cincinnatus, about 2,250 years earlier, having won against the fierce Aequians as a Roman general, refused absolute power and instead went back to his farm and his plow. One of Washington's generals, the redoubtable and later secretary of war Henry Knox, formed the Society of the Cincinnati in 1783, comprising Washington's commanders in the field from the Revolution and their descendants. The Society exists today and has a national office in Washington, D.C.

WASHINGTON DID NOT JUST ARRIVE WITHOUT CAUSE AT HIS EXALTED STA-tus, beloved by his fellow countrymen for over two hundred years, with children, cities—including Washington, D.C., in 1791—monuments, mountains, schools, even states and holidays named in his honor. And later, as the standard by which all future presidents would be measured, recorded as also the greatest president by most historians—a man who would be widely revered for his integrity, grace, manners, charm, Christian faith, and humility. His devout mother played a key role in the development of his character. While he was sometimes described as having little genuine affection for Mary, the reserved Washington still credited her with his principled and moral upbringing. Indeed, this was inevitable, for when George was eleven, his father died, leaving Mary Washington a single mother.

"The relationship with his own mother was laden with difficulty for both of them. Self-centered and acquisitive, Mary Ball Washington was preoccupied with her eldest son to the virtual exclusion of her other [four] children. That preoccupation expressed itself in fears for George's safety, pleas not to put himself at risk in military action, and demands for assistance, usually monetary, even though she continued to occupy and enjoy the profits of his property on the Rappahannock," said Patricia Brady in *Martha Washington: An American Life.*[2] Another book described her as "cantankerous and demanding."[3]

HIS FATHER, AUGUSTINE ("GUS"), HAD A "NOBLE APPEARANCE AND MANLY proportions."[4] As a youngster, George traveled far and wide visiting friends and relatives, and he was away when he learned that his father was dying. He returned immediately to Ferry Farm on April 12, 1743, the day his father died—"His father was stretched out on a bed pending his burial. He may have been an absentee parent, but his sudden loss left an emptiness in George's life, a vacuum that needed to be filled."[5] George was just eleven years old. Augustine's will was shortly probated and he left his son George the plantation of Ferry Farm, an equal division of slaves and estate, and a lot in the newly formed and nearby town of Fredericksburg. Mary was to supervise the children until they came of age.

George's fourteen-years-older half brother, Lawrence, went off to fight in the amusingly named War of Jenkins' Ear in 1740, "leaving George consumed by his loneliness." When Lawrence returned, to the admiration and hero worship of his younger half brother, he settled fifty miles north from the family, away at a plantation named Epsewasson, high above the Potomac River. Lawrence later renamed it Mount Vernon, after the famed British admiral Edward Vernon, whom he deeply admired.[6]

Mary Washington, née Ball, Augustine's second wife, lived at Ferry Farm and raised George, tutoring him, admonishing him, driving him to distraction often, loving him, and also fashioning the boy who would eventually become one of the greatest men in history. One historian wrote of young George's and his mother's strength of wills as "incompatible," but this, said Dorothy Twohig, "fostered young Washington's independence and self-reliance."[7] Later, control of Ferry Farm, which Augustine had left George, became a source of irritation between the son and his mother.

MARY WAS BORN AROUND 1708 OR SO—THE EXACT DATE IS NOT KNOWN. (Many ages in this book are approximated.) Much of her life was a mystery, sometimes placing her in some historical studies as less of a person and more of a mythic figure. Her family, the Balls, were prominent in the Millenbeck and Epping Forest parts of Virginia's Northern Neck, jutting out into the Chesapeake Bay, adjoining the Potomac on one side and the Rappahannock River on the other.

Though she was well provided for by her husband, it was tough going for Mary Ball Washington after Augustine's passing. She was a widow in her late thirties, raising six children all by herself, supervising farms, supervising slaves, supervising her family. She never remarried, though she would have been an attractive catch, at a still desirable age and very wealthy. But it was well known around Fredericksburg, Virginia, that she was a handful and at times frustrating. Her son, by contrast, was equally legendary for his reserve.

However, his resolute reserve in adulthood had not always been so. In his youth, George Washington had a fearsome temper. This was observed not just by his mother, but by people such as Thomas, Lord Fairfax, in a note to Mary: "I wish I could say that he governs his temper. . . . He is subject to attacks of anger on provocation, sometimes without just cause."[8] Perhaps he learned something from his mother.

George's capacity for rebellion may have been prompted by Mary's overprotective care. His later language describing the Revolution evokes a mother and son. The Mother Country, George wrote, "thought it was only to hold up the rod, and all would be hush!"[9] and Great Britain "did not comprehend America—She meant . . . to drive America into rebellion, that her own purposes might be more fully answered."[10]

Through the following decades, from adolescence to adulthood, Mary was a near-constant irritant to the maturing George. Financial demands were but one of the many issues plaguing the relationship between mother and son, though her supposed loyalty to the English Crown during the most important chapter of American history may have also angered him. She was probably a royalist, at least initially. "Mary Ball Washington . . . never expressed support for the Colonies' cause," declared one historian, Bonnie Angelo. In this fashion, she was like another revolutionary's mother, Jane Randolph Jefferson (who was also skeptical of independence for the colonies), though her son Thomas cowrote the most important revolutionary document in history.[11]

But in a hotbed of revolutionary fervor such as Fredericksburg, where munitions were made for the colonies by Mary's son-in-law Fielding Lewis, this would have been cause for concern—risking ostracization and being outcast—except that her son was leading the Revolution, and this protected Mary from any acts the local citizens might have contemplated taking against her. At least none were reported or recorded at the time, and she continued to be seen around town—shopping, going to church—untouched.

IN POST–REVOLUTIONARY WAR VIRGINIA, MANY THOUGHT MARY, NOW AT AN increasing age, could no longer live at her Fredericksburg home. One of the many included George, who wrote to his mother urging she leave and choose one of her children to live with—though not him. He wrote, "Candour requires me to say" that moving to Mount Vernon (which he'd owned since 1761) "will never answer your purposes, in any shape whatsoever—for in truth it may be compared to a well resorted tavern, as scarcely any strangers who are going from north to south, or from south to north do not spend a day or two at it." He continued, "Nor indeed could you be retired in any room in my house; for what with the sitting up of Company; the noise and bustle of servants—and many other things you would not be able to enjoy that calmness and serenity of mind, which in my opinion you ought now to prefer to every other consideration in life."[12] Washington was clearly giving his mother unsolicited advice while also telling her there was no room in the inn, in this case, the Mount Vernon inn, his home. In this matter, she took his counsel and never moved to Mount Vernon.

Two years later, Mary died around the ancient age of eighty, in late August 1789, at her home in Fredericksburg, the town near where she lived for a majority of her life. Through these decades, she saw her eldest son become almost a god to many American citizens. He addressed her mostly with respect in his few letters to her, no matter his personal feelings. Through anger and resentment, financial worry, and updates on her health and well-being in times of war, George always addressed his letters to his mother as "Honored Madam." This salutation served two purposes: it showed his respect for her while holding her at arm's length.

WITH MARY BALL WASHINGTON, THE IMAGE IS A MIX OF THE UNKNOWN AND whispers of history, the facts and the legends. She was the real mother of George Washington; she certainly existed; yet as part of the mythos and mythology of the era of the Founding Fathers, of revolution and renewal, much has been lost to the ether of history. Was she part helicopter mother, part "Mommie Dearest," or was she a saint and a joy for George? Historians through the years have portrayed her as both.

Mary Ball Washington remains often misunderstood. She does appear to have been a thorn in George's side ever since he was a teenage boy—but also a cherished rose.

There is frankly a lot of conflicting information on George's feelings about his mother, even as an adult. "Added years and understanding brought no improvement in his relations with her. . . . Apparently he did not write her even once during the war," wrote eminent biographer Douglas Southall Freeman. Although this comment is in dispute, as it had been previously reported that she did get at least one piece of correspondence from her son during the Revolutionary War. Others may have been lost by Mary or from George. The same historian noted that Washington, while possessing "magnanimity and patience in dealing with human frailty," unlike his mother, did not extend such grace *to* his mother, as "he felt she had been grasping and unreasonable."[13]

Many histories have tended to focus on only one or the other side of Mary's character. She is seen as a saint or a villain, nothing in between. But in fact, taken together, Mary's seemingly contradictory character traits complemented each other. Mary's kindness and control were one and the same. Mary Washington was a woman who used a facade of motherly virtue to cover her desire to control her son. In the same way that he led a country to break away from its overbearing imperial matron, George had to struggle to find independence in his own life, to step away from the power of his demanding mother. It was a struggle that lasted all her life. Even after Mary died, George was left dealing with her tangled affairs.

THIS BOOK IS JUST AS MUCH A HISTORIOGRAPHY OF MARY WASHINGTON AS IT is a biography. Very few correspondences exist about this important woman. It was also the nature of the period that it often placed her on the sidelines, being a woman and not being married to a famous man. Opinions of the relationship between Washington and his mother vary. There were a number of hagiographies written in the 1800s and early 1900s such as *The Mother of Washington and Her Times* and *The Story of Mary Washington*, both by women. The latter of these books was reviewed by the *American Monthly Magazine* as "a warm welcome" for those interested in early American history.[14] Indeed, while it was certainly a delayed canonization compared to her son's, some early biographers of the mother of Washington—it sounded saintly, like the Mother of Christ, with no more qualifiers needed—almost painted her as a goddess. "Roman matron" came up sometimes, an appropriate title in light of the Cincinnatus George. One called her "the 'Hero' to a Spartan matron" and further a "Christian Matron."[15] These books and many more from that era had her as less a woman and more a deity who could do no wrong.

But slowly, a new image of Mary Ball Washington began to emerge in the opposite direction, aided by some harsh letters from Washington himself to his own mother that were cataloged and widely studied. Very few exist, but those that do paint a nasty relationship. This image of Mary in historical circles shifted entirely from whispers to shouts with the publication, between 1948 and 1957, of the massive and unyielding seven-volume biography of the first president, from cradle to death, by Douglas Southall Freeman. In an about-face of how previous biographers treated her, Freeman, in the first volume, called her a "mistress of much or of little, mistress she was resolved to be."[16] It wasn't protection that she gave her eldest son, who at one point in his youth wanted to risk everything for a dangerous life at sea; it was *authority*.

Describing Mary, George's cousin Lawrence said, "I was ten times more afraid [of her] than I ever was of my own parents; she awed me in the midst of her kindness." That kindness, and Lawrence did call her "truly kind," was subsumed in Mary.[17]

As a teenager, young George had wanted to join the British

navy as a cabin boy, so Mary wrote a letter to her half brother, Joseph, in London, asking his advice. Within a short period of time, he wrote back, telling her that under no circumstances could George, an American, be allowed to join, where he would be treated like a "dog." A caste system existed in the British navy, as it did throughout the English culture. First came the British royalty, then British subjects, and finally, at the end of the list, came Americans. It was also a time when a substantial number of cabin boys died at sea—they developed scurvy, they were washed overboard, they were killed in battle. It was a risky and dangerous existence for young boys. While she had previously approved the venture, she vetoed it now. It was the first of several attempts to keep George from entering dangerous undertakings.

James Thomas Flexner's four-volume biography of George from the 1960s and early 1970s, also unmistakable for its importance in Washington scholarship, derided Mary's overcontrolling parenting. In his romantic teenage boyhood, George was in love with Sally Fairfax, a young woman two years his elder. When he was sixteen, Flexner said, "he was still, however resentfully, tied to his mother's apron strings, in a rundown farm, full of younger children, which he would have undoubtedly been ashamed to show the elegant Sally."[18]

Thus the fickleness of the history of Mary Ball Washington.

It should also be noted that authors, from both the history and hagiographical sides of the debate, when approaching Mary Ball Washington, sometimes simply got facts wrong. Flexner, for instance, mistakenly called George's only sister, Betty Washington Lewis, "Mary Lewis."[19] A slip of the finger, sure, that is bound to happen in such massive works. On the other hand, Lydia Sigourney, in a small book published in 1866 for schoolgirls about powerful women through history, said that Mary died after George, over a decade in error.[20]

The immeasurably important relative and descendant George Washington Parke Custis sometimes also got chronology wrong in Mary's and George's lives. Others simply wrote thinly sourced stories with contradictory facts; for instance, how Mary and her husband, Augustine, met in England when the latter was tended to by the former after an injury, like some story seen in a romance novel. All these works have shaped the legacy and portrait of who Mary Ball Washington was—and was not.

But Flexner's and Freeman's works are, without fail, some of the most comprehensive biographies of the first president's life, as is Ron Chernow's masterpiece.

Slants or presentism in history, whether consciously or not, encouraged the reader to ignore context or brush away any excuse that may contradict it. What Mary or others may have seen as protective in George Washington's teenage years, Freeman sees as controlling and authoritarian. This bias erases facts, maybe implying something that it is not. This is dangerous.

This book is just as much a history of Mary's times as her near century of life. The eighteenth century in the New World saw a fundamental change that echoed across the globe, as the loyal colonies, fighting for the Crown in the 1750s, went against King George III twenty years later, establishing a system of government previously unseen in history. Whether she knew it or cared for it or supported it or not, how Fredericksburg was run, how the colonies fought, who died nearby, and how nations were forged affected her. Though she probably knew little of it, the way in which the Continental Congress in 1781 debated the ins and outs of the Articles of Confederation and how it again debated a new constitution in 1787 affected her. While she was a key influence on one of history's greatest men, during the Revolution she was but one person in a vast system known as the American experiment.

THIS "HONORED MADAM" WHO DIED NEARLY THREE HUNDRED YEARS AGO was many things, often contradictory and often paradoxical: we know many things of her, and we know little; we know how she treated her son, but we have no known motive; we have interpretations of historians over centuries, both sides getting some basic known facts wrong, which makes the modern reader question what else was wrong in these fundamental works. But despite their difficult relationship, George and Mary shared an important bond of blood. "Whoever has seen that awe-inspiring air and manner so characteristic in the Father of his Country, will remember the matron as she appeared when the presiding genius of her well-ordered household, commanding and being obeyed,"[21] wrote George's cousin. And as George himself wrote, it was by Mary's "Maternal hand [that,] (early deprived of a Father) I was led to Manhood."[22]

She was the mother of the Father of our Country. She was authoritarian and frustratingly singular, the royalist woman who shaped history's most famous patriot.

She was Mary Ball Washington.

Chapter 1

Virgin Land and Virgin Love

LIFE IN AND AROUND THE COLONY OF VIRGINIA

1600–1780

"The specter of the Old World . . ."

In 1651, the county of Lancaster, Virginia, was created out of a combination of York and Northumberland counties. A few years later, George Washington's ancestors arrived in the New World and made their way here, where Mary Ball Washington was born in 1708.

By the end of the Revolutionary War and the independence of the American colonies from England in 1783, the number of Virginia's counties tripled as population shifted and frontiers opened. Spotsylvania County, established in 1721, was a prominent place of interest for the Washington clan—and it was in the county's largest city, Fredericksburg, that George was raised into adulthood, and where Mary lived for most of her life.[1]

Slowly but surely the New World became less a temporary way of life and more a permanent home. New settlements were springing up, port towns like Alexandria were in the proto-stages, new grants for businesses were signed, and new manors were built. Iron ore, useful for making tools, weapons, structures, and more, was discovered here and became particularly important for settlers, especially as a means of cutting off English dependency on Sweden.[2] John Carter had built the first church in the region.[3] His son, Robert Carter, born 1662 on Corotoman Plantation in

Lancaster County, acquired enough land and wealth appropriate for his nickname, "King" Carter.

THE SPECTER OF THE OLD WORLD WAS NOT YET BANISHED. ONE OF THE most prominent houses in Essex County was Ben Lomond, a Georgian manor a short riding distance from the town of Tappahannock. With land stretching out to the Rappahannock River, it saw frequent renovations through the years as prominent Essex family members settled in, including its most famous occupant, Judge Muscoe Garnett. It was named after a mountain in the Scottish Highlands.

IN THE NEWLY CHARTERED PRINCE WILLIAM COUNTY AROUND 1740, THE brick home of Bel Air was erected by Captain Charles Ewell, whose wife, Sarah Ball Ewell, was a younger cousin of Mary Ball Washington. (Decades later, Charles Ewell's granddaughter Frances and her husband, Mason Locke Weems, moved into Bel Air, from which Weems would write the first posthumous and legendary biography of George Washington in 1800.)[4]

The Northern Neck of Virginia, the geographically northern peninsula that stretches from the Potomac to the Rappahannock, was bustling. Prominent families like the Carters, the Lees, the Washingtons, the Balls, and the Wares all settled in this area, quickly making it more than just another remote region of the colonies.

It was beginning to look like home. People were no longer looking to Europe, the Old World, for guidance and sustenance. They had become Americans.

IN THE PAPERS AND DOCUMENTS OF THE DAY, THOSE WHO CAME TO AMERICA from the Old World were routinely referred to as "immigrants," while Native Americans were frequently referred to as "hostile Indians." "The early pioneers found the Indians very troublesome," wrote F. L. Brockett and George Rock in the late nineteenth century, "and in order to [secure] the protection of their families from the raids of these unwelcome visitors, they were compelled to almost constantly be under arms, as these *visits* were generally made when civilized people were supposed to be asleep."[5] This was an unceasing issue into the eighteenth century, as white settlers continued to expand into former Indian territory. Mary herself heard of these stories, akin to boogeymen or ghost stories by campfires, of entire villages and families being killed. There was some truth to these stories, as her ancestors well knew of the scathing relations between Native Americans and colonizing Europeans. It was a looming threat.

Things were certainly better in Mary's time than they were many years earlier in Jamestown, down the Potomac River and out into the Chesapeake Bay, where Captain John Smith's establishment failed and the residents fell to cannibalism during the "starving time." Later reports were macabre and ghastly, including one of a husband who killed his wife and then salted her flesh. Eighty percent of the colonists died, having starved to death or being killed to eat. The population dwindled from three hundred inhabitants to sixty within months.[6] Such haunting stories reverberated throughout the colonies and indeed back to England. This terror from America may have easily struck the attention of William Shakespeare in his play *The Tempest*, whose inspiration of islands and dangers unknown to the average Londoner may have been based partly in truth.

Virginia's population increased; so did the means of communication. The *Virginia Gazette*, founded in Williamsburg in 1736, offered weekly updates from across the royal colony. It was essentially a statewide paper. Various competitors emerged in the latter half of the paper's life, before its final publication in 1780. Regional newspapers started popping up, focusing on local news— everything from advertisements of land and business, to notices of stolen property and runaway slaves, to major news in England or Europe. Breaking news took precedence, though by the time a crisis was known to the printer, it was many days old. The same applied to courier letters.

IN THIS ERA, TOBACCO WAS THE FAVORITE CROP, BUT CORN, RICE, AND OTH-
ers also found room to thrive. Whole armies of freemen and slaves
were devoted to the growing, harvesting, curing, hogsheading,
and shipping to Europe of the much sought-after tobacco leaf.
Currency was often tobacco weighed by the pound, as an alternative to the pound sterling in England. Spanish coins received
from trade were sometimes used, but tobacco was the most stable,
and more local than the far-off English or Spanish coins. Corn
and wheat were also accepted to pay off debts. In 1751, over a
hundred thousand pounds of tobacco was accepted to help build a
church in Stafford County. Clergy was also paid with tobacco, and
merchants and prospective buyers bought anything, from ferry
passages to books and land, with the crop.[7]

Credit was another popular method of purchasing goods and
services, and a relatively new one at that. "Credit refers not just
to the money you possess but also to what others believe you will
be able to pay. One economist has described it as 'money of the
mind,'" as historian Joseph Ellis described the colonial novelty.
It was risky, but it was a step necessary for building the economy that would be known as *capitalism*.[8] The prevailing economy, though, was mercantilism, which held the belief of a static
amount of capital in the world, and that through economic activity, if one got more, it meant that someone got less.

IT IS WRONG TO THINK THAT WOMEN DID NOT ENGAGE IN PUBLIC AFFAIRS OR
commerce in the 1700s. Names like Elizabeth Burtin and Ruth
Day and Sara Knowles and Susan Phelps and Ruth Smily were
frequently in the papers, organizing Alexandria, Virginia, as early
as 1669.[9] They also included Mary Ball Washington. After her
husband's death, she was engaged not only in child-rearing as a
single mother but also in supervising large tracts of land, farms,
and slaves, the buying and selling of horses, and crop management and sales.

To be clear, it was still a male-dominated world, but women
had plenty of important roles to play as well. Often, as in Mary's

case, they had their motherly duties of raising children, but they also were granted responsibilities based on their husband's or father's line of work. Once slaves became more common, women's labor on the farms shrunk to labor in the house.[10]

There were instances of women taking more active roles in their communities, such as becoming ferry keepers, jailers, preachers, shipwrights, spinners, and poets. Still, Abigail Adams felt the need to remind her husband, famously, to "Remember the Ladies, and be more generous and favourable to them than your ancestors" when John Adams was to draft laws for the new country.[11]

FARMING WAS THE PRIMARY INDUSTRY IN THE COLONIES IN THE 1700S. "IN the backcountry or in the districts remote from transportation, a man could buy a large tract for little money and often on credit. Going into the forest with his family, the frontier farmer built a crude shelter and began to cut down the trees, usually at first merely girdling the larger ones so they would die and let in enough sun to grow corn between the stumps. His efforts the first year or two were devoted to survival, raising enough food for subsistence."[12]

One observer wrote that American farmers were somewhat similar to Native Americans, eking out a marginal living, though the Americans could also operate as "their own carpenters and smiths." Still, it was possible for American farmers to eventually prosper, to "in [a] few years maintain themselves and families comfortably."[13] They could in time raise themselves up and join the local society, without regard to station or church. Another writer noted that America was "one of the best places in the world for a poor man."[14]

Contrary to a popular conception of America as a vast, empty new world, land was not always widely available in eastern Virginia, so many opted for other careers. "In most villages . . . blacksmithing, milling, weaving, brickmaking, keeping a store, and sometimes operating an iron mine" were all options.[15]

THE ROMANTIC, SOFT IMAGES OF COLONIAL AMERICA PORTRAYED BY BOOKS or movies in later centuries were fabrications at best, even in the rural areas. Life was often dirty and diseased in colonial America in the 1700s.

In the summers, flies and mosquitoes were unbearable. John Smith wrote tales of Native Americans whose "bodies are all painted red, to keepe away the biting of mosquitos. They goe all naked without covering."[16] Bees and even the common housefly were found in Virginia, introduced by settlers. Thomas Jefferson, in his *Notes on the State of Virginia*, wrote in the late 1700s that Native Americans "call them the white man's fly, consider their approach as indicating the approach of the settlements of the whites."[17] Disease and contagion were a constant threat, with influenza and smallpox wiping out settlements. Personal hygiene was reserved for the upper classes and, even then, infrequently availed upon. It was long before the time of Louis Pasteur and his germ theory of disease.

THERE WAS A VIBRANT LOWER TO MIDDLE CLASS IN THE MIDDLE COLONIES. "I have found them having for dinner potatoes, bacon, and buckwheat cakes," observed one Frenchman. "For downstairs rooms, a kitchen and a large room with the farmer's bed and cradle, and where the family stays all the time; apples and pears drying on the stove, a bad little mirror, a walnut bureau—a table—sometimes a clock; on the second floor, tiny little rooms where the family sleeps on pallets, with curtains, without furniture."[18]

There was also a small but distinct upper class that was "bound to one another, and in some colonies to the landholding gentry, by intermarriage, mutual patronage, and common business interests. They were universally recognized as 'the better sort.'" Mary Ball Washington was part of the "better sort," as the Washington clan was in the upper gentry. Opinions were diverse as to whether she acted like it for a widowed woman of her status. The better sort was sometimes derided as "turtle eaters."[19]

It was a time of social mobility. Shopkeepers were already

prominent in colonial America, and the marketplace was open for boys to become apprenticed and then adult craftsmen. Woodworking, barrel making, leather tanning, cloth making, tailoring, shoemaking—all were possible paths for the apprentice boy. Indeed, Thomas Paine, the conscience of the Revolution, was a corset maker when not calling upon his fellow colonists to use common sense. Up in Massachusetts, the great Paul Revere was a silversmith, famous for his large bells.

MARY PROVOKED REMARKS LATE IN LIFE BY THE STUBBORN SIMPLICITY OF her dress. It was typical of her, however, to defy the fashion and forge her own path. It was also very American.

Such was not the case in her childhood, however. Though the Americans would have naturally been behind in European fashion with the slow speed of cargo ships, stylish clothing was still very much in vogue. The type and quality of clothing were two indicators of wealth. For the rich it was layers of complicated and ornamental dress. "The petticoats of sarcenet [had] black, broad lace printed on the bottom and before; the flowered satin and plain satin . . . with rich lace at the bottom."[20] Clothing was exported throughout the world and into Virginia. Manufacturers advertised in the *Gazette* any chance they got, and clothing was often for sale.

Among one of the fashion trends during the eighteenth century was the macaroni, outlandish and over-the-top fabrics in bright colors to signify travel experience in Europe. The poem "Yankee Doodle" of around 1755 mocked the American colonists' bastardization of the trend, thinking that a simple "feather in his hat" was enough to be called fashion.[21]

The women "adorned themselves in English fashions and [spent] much time visiting back and forth over their country estates. Lavish balls and dinner parties became the subject of public notoriety, arousing the indignation of the lower classes." Even there, class envy and class warfare were both present. Women shopped daily, as there was no refrigeration in homes, to the butcher and baker and greengrocer.[22] Said Anglican minister Charles Woodmason, visiting the New World from Europe: "The mother looking as young as the daughter. . . . The men with only a thin Shirt and pair of Breeches or Trousers on. . . . The women bareheaded, barelegged and barefoot with only a thin Shift and under Petticoat."[23]

As the Revolution began, American trends tended to become more simplistic. Men's wigs, which could be worn by any aged adult man, became less outrageous. Women's hair and clothing shifted from imported English goods and styles to homemade and modest looks. It was a sign of patriotism to renounce the provocative English fashion that came from overseas. "Of course simplicity of dress was noticeable," wrote Dorothy Dudley in her diary in 1776. "No jewels or costly ornaments—though tasteful gowns, daintily trimmed by their owner's own fingers, were numerous." Domestic products were in; English royalist fashions were out.[24]

SLAVERY WAS VERY MUCH A PART OF THE CULTURE AND BACKGROUND OF colonial America, especially in the South, where staple crops like tobacco needed tending. At first, white indentured men and women from Europe, and later, Native Americans, toiled on the land, but indentured servants eventually worked off their status. As the indentured became "un-indentured" and developed their own small farms, the need grew for reliable, durable, and cheap (free) labor.

After 1619—when Great Britain entered into the slave trade— some ten million African slaves were shipped to North America and South America, all the way down to Brazil. "In Virginia, during the 1680s, there were still only 3,000 slaves in a population of 70,000, but by 1756 they numbered over 100,000, about 40 percent of the population," wrote historian James Ferguson.[25] There was not just an increase of supply but an increase in demand for slaves. In agriculture, in housework, and in entertainment, "negros" or "mulattos" were sold and bargained for. One man, William Fearson of Williamsburg, asked for "an orderly Negro or

Mulatto man, who can play well on the violin." If one were to give or rent away their slave to Fearson, he noted, the master (and not the slave) "may have good wages."[26] Other slaves were carpenters or barbers.[27]

There were of course runaways, which came up in every weekly newspaper. Sometimes details were provided, physical descriptions of brands or height or facial features. Sometimes these included when the individuals were last seen. All demanded they be returned for monetary reward, and one can only imagine the horror stories behind closed doors—all legal, of course, since the 1660s. "Stringent racial laws" hit the Virginia lawbooks, which were "designed to regulate white-black relations and provide planters with greater powers to discipline their slaves." This in turn led to a greater need for slaves, as slaveowners saw them more and more as expendable property.[28] One influential planter, William Byrd II, was a known abuser of slaves, even to the shock of eighteenth-century Virginia. In his diaries, as normal and mundane as saying his prayers or having breakfast, he noted the numerous times he whipped his slaves. One slave was forced to drink his own urine after wetting the bed. A certain "Jenny" was whipped for being the "whore" of another.[29] Rebellions and uprisings occasionally occurred, but for the most part, slaves, over the years, unhappily adapted to their lot. A few times, conspiracies involving whites and blacks plotting mutinies were discovered but they were quickly dealt with, usually resulting in the hanging of members of both races.[30]

There were emancipators and abolitionists, George Washington among them, who viewed slavery more as a necessary evil for society. In 1790, several Quakers petitioned the United States Congress to abolish the slave trade and, ultimately, to abolish slavery itself. "Because they did not believe that God discriminated between blacks and whites," wrote historian Ron Chernow, "many Quakers had freed their own slaves and even, in some cases, compensated them for past injustices."[31] But the institution continued, unabated.

Still, slaves and slavery as an institution created no moral dilemma in many plantation owners' minds. They were an essential
part of their lives. They were as essential as the crops they har-
vested or the livestock they raised. Announcements of their trade
were as frequent and mundane as the sale of land.

Besides the abuse or neglect that slaves received, besides the fact
that their emancipation would not come for over a century, they
adapted. Some attended religious services, though one Anglican
minister noted that they "cannot understand my language nor I
theirs." There was one story of an unnamed slave to an unnamed
master in Virginia, whose relationship to an English indentured servant created a hybrid language. With a mix of an African dialect and the servant's native Warwickshire English, the
slave learned words like "his'n," "her'n," "howsomdever," "yarbs,"
"pearts," "ooman," and other slang from central England.[32]

WHILE MARY'S PEDIGREE WAS AS COMMON AS THE VIRGINIAN SOIL, HER PO-
sition in society was the closest thing America had to an aristocracy. As the colonies evolved, a "planter elite" began to emerge
in the South. "The planter elite was the nearest thing to a European aristocracy that America produced." However, unlike in
the royal elites of Europe and its strict caste system, including
the notion of divinity, there was social mobility in the American
planter elite. One could, through the sweat of his brow, rise up
into the new planter class—but also just as easily fall out of it.[33]
As the planter culture evolved, so did entertainment among the
upper classes. These big plantations were often far from cities,
towns, and even other plantations; hence, parties and entertain-
ing often lasted a week, with guests staying for extended periods
of time. Said historian Ferguson, "Plantation mansions, built to
accommodate many guests, were the scene of frequent parties,
which sometimes stretched out to over a week of lavish entertainment. Both sexes studied conversation as an art and cultivated
their ability to dance. Gentlemen gambled at cards and horse racing and rode to the hounds after the English manner. Beneath the

apparent frivolity and leisure of plantation life was a utilitarian core of duties involved in administering its economic and human activities, but the outward face of planter culture was aristocratic in its dedication to luxury and gracious living."[34] Weddings were just as much a time to socialize as a time to wed a couple, with no honeymoon. Instead of the newlyweds going to faraway paradise islands, they and the guests would often stay overnight in the same house.

MARY WAS A MEMBER OF THE ESTABLISHMENT RELIGION—ANGLICANISM—IN a period full of new denominational diversity. Slowly, the Great Awakening in the colonies—a spiritual movement—united the Carolinian with the New Yorker. It had started in the 1720s, in England, and quickly spread through the eighteenth century to the colonies.[35] It was decidedly anti-elitist and pro-populist, the notion that "emotion, not doctrine, was the crux of its message." It elaborated that religion "was a matter of the heart, not the head, and that simple, ordinary men were more likely to be touched by the divine spirit than were the sophisticated upper class or an educated clergy which was dead to spiritual values." The Great Awakening was emblematic of the colonies themselves and their attitude toward Europe generally, Great Britain specifically, and most authority at the time. Indeed, it was, according to Ferguson, "anti-intellectual in spirit and rejective of authority, preaching the spiritual worth of the common man."[36]

THE ENLIGHTENMENT'S VERY NAME MATCHED THE MOVEMENT. INTELLEC-tuals saw the status quo of absolute monarchy, absolute authoritative religion, and an absolute caste system, and rejected it all. John Locke's "Essay Concerning Human Understanding" in 1689 rewrote what it meant to be human. Science and reason were placed on a pedestal, rejecting religion and the superstitions of old. Isaac Newton, Benjamin Franklin, and Voltaire—all names that are easily recognizable today and were as famous in their own day—spoke out.

Among the new "religions" coming from the Enlightenment was deism, a belief of reason over the unexplained, a "natural religion" versus a supernatural religion, where God created the world and played a passive role in its everyday occurrences. There is a Creator, but he did not stick by for miracles or divine intervention, they believed. As such, it directly contradicted essential Christian tenets from all denominations, including the divinity of Jesus Christ and his resurrection.[37]

That put it at odds with the established Church of England.

As English colonies under English law and an English parliament, the official state religion of Anglicanism—literally, "English-ism"—recognized the sovereignty not just of the Crown but of the Church. It was a distinction going back centuries to the days of Tudor king Henry VIII, who decided that the pope of Rome was not to have any authority over his native church. In the pope's stead, Henry placed himself as the head of the Church, the supreme authority and Defender of the Faith.

Before 1776, the dominant religion in America was the Church of England. From tithes—compulsory taxes that paid for ministers' salaries and the construction of new churches—to its official status as the colonial religion, the Anglican Church was the spiritual arm of the Crown. Advertisements were frequently nailed to church doors on the latest news of the day, making the church structure not just spiritually but practically necessary.[38]

It was not all-powerful, though, which the colonists would come to realize in the late eighteenth century. There was no separate bishop of the colonies; they fell under the jurisdiction of the Bishop of London. With such far-off leaders in both the episcopacy and royalty, some of the Anglican churches took a distinctly American tone as decades moved forward. "The colonial church," said Mary Thompson, historian at Mount Vernon, "saw its purpose as a practical one of providing spiritual sustenance and teaching to individual members, as opposed to a more intellectual involvement in theology and philosophical theorizing." Obedience to God's law—and thus the church and the state—was taught above all.[39]

But still, despite the laws in place mandating compulsory Sunday worship, the practicality of rural Virginia "meant that church attendance, while important, was emphasized less than private devotions, which could be done at home."[40] It was not a rejection of the clerical arm of the Church to worship at home, but the sheer inability of the already low number of ministers to cover the colony that made it necessary. Thus, formal catechism classes did not exist, instead moving to the home plantation under the thumb of the parents, where biblical and moral lessons were taught alongside a secular education.

CATHOLICISM WAS ANOTHER MATTER. THERE WEREN'T MANY ADHERENTS in the colonies at the time—maybe a few thousand, mostly in Maryland. Chesapeake Bay Catholics would have found relative sanctuary in Maryland. The first Mass in the thirteen colonies was said on the shore of Maryland by Father Andrew White, Society of Jesus, in 1634. King Charles I of England had previously given George Calvert, a Catholic convert, a land grant north of Virginia, a charter that continued under his son Cecilius. Still, anti-Catholicism took hold in Maryland, leading to the repeal of the Act of Toleration in 1654 and the death sentence of ten Catholics (of which four were carried out), and the destruction of Jesuit property. Soon after, the Church of England was established as the only official church of the colony, and Catholics were forced into hiding.[41] There would be no public Catholic churches until the repeal of the penal laws in 1776; within a couple of decades, there were hubs of Catholic churches such as Saint John the Evangelist in Frederick, Maryland; Saint Mary's Parish in then Barnstown (now Barnesville) in 1807; and Mount Saint Mary's Seminary in 1808—all founded by the same Father John Dubois, an asylum seeker fleeing the violent and anticlerical French Revolution.

Other colonies with other predominant religions placed more emphasis on doctrine, often harshly. The capital crimes of Massachusetts in 1660 listed anyone who "shall have or worship any other god, but the lord god . . . [to] be put to death." Puritanism had a much stronger hold in the North than in the other states. If anyone was accused of being a witch, or "blaspheme the name of god, the father, Sonne or Holie ghost," he would be executed. Murder, bestiality, sodomy, adultery, theft, lying, and rebellion were all capital crimes, listed after the religious crimes.[42]

Virginia was also intolerant. Lancaster County in particular was more severe than other Virginian counties. In 1685, John Chilton was fined five guineas for profanity on a Sunday; and several women were publicly whipped, such as one unnamed woman who received eighteen lashes for fortune-telling. She was afterward ordered to attend the next religious service at her parish, "draped in a white sheet with a wand in her hand to beg for forgiveness of some person she has slandered." For capital crimes, Lancaster County was again blinkered. Before their death by hanging, the sentenced were given new clothes and a hearty breakfast of chicken, fried oysters, cakes, and coffee.[43]

NOT ALL WAS AWFUL IN RURAL VIRGINIA IN THE 1700S. AMONG THE FAVORite activities were sporting and horse racing and equestrianism, the last of which the young maiden Mary Ball was said to have loved. Horses were of course necessary for transportation and they were a sign of wealth, but horse racing was important, and the *Virginia Gazette* was full of advertisements for and the results of these races. These were often major social functions. Horses were either bred or imported from England, often the best of the best; only the wealthy could afford them.[44] It was very much a holdover of European society that continued even through the Revolution.

Taverns, too, were hot spots of gossip and socializing. These were primarily a man's place, with women—usually the wives—segregated to a smaller room. In every major and minor town throughout the colonies, taverns were safe havens and overnight stops for travelers. They were in every shape and size depending on need, with multiple tavern boys and tavern wenches assisting. One tavern near River Road between Richmond and Williamsburg, advertised for rent, was described as "commodious, having four Rooms below and two above Stairs. There are Outhouses, Garden, Pasture, and Land also, if required, sufficient to work two or three Hands."[45] At others, it was common for guests to dispose of lice and other parasites.

SUCH WAS THE WORLD OF MARY BALL WASHINGTON, HER HUSBAND, HER AN-cestors, her descendants. Such was colonial Virginia, making up the totality of Mary's life and the majority of her son George's. Such was the world, predominantly run by men, in which a woman, named for her mother, was born and raised, whose oldest son eventually changed not just the land on which she lived but the world.

Chapter 2

"To Look to the Sky"

GEORGE WASHINGTON'S GENEALOGY

1600–1708

"The word Ball means bold, courageous. . . ."

N ot much is known of Mary Ball's ancestors," wrote Charles Moore in his 1926 work *The Family Life of George Washington.*[1] Another source admitted, "Little is known of her ancestors, except she inherited an unimpeachable name."[2] A greater emphasis over the years had been placed mainly on the paternal ancestors of our nation's first president, leaving much of the maternal studies to the wayside and unknown.[3]

But that does not mean there was nothing.

George Washington's earliest maternal relative, some speculate, was Drogo de Montacute, a close ally of King William the Conqueror, formerly known as Duke William the Bastard of Normandy. William invaded the Kingdom of England as one of the contested heirs to the throne upon the death of King Edward the Confessor in 1066.

IT'S POSSIBLE TO TRACE THE EVOLUTION OF THE BALL NAME THROUGH THE years. "Ball" was originally "Baldwin." According to Earl Leon Werley Heck, genealogist and historian, "The surname Ball, according to the best authorities, dates from Norman times and is a shortened form of Baldwin, which family were for many generations Counts of Flanders."[4] He pointed to two sources for this:

William Camden of the sixteenth century and Charles Bardsley, both of whom give an etymological origin of Baldwin to mean "bold victor."[5]

Through the centuries, the Baldwins eventually became the Bals (the first of these appeared in the thirteenth century with landowner Vice-Comes Bal[6]), the Bales, the Balles, the Baells, each etymologically different from the other—one is pure Saxon, another Anglo-Saxon, some Norman. "We find the Baell corresponds to the Anglo-Saxon word Báel, which means a funeral pile; and Ball approaches the Saxon word Báld, denoting bold, audacious . . . Ball is a diminutive of Baldwin, which latter name is an Anglo-Saxon word, meaning bold in battle (win-battle). I am of the opinion," said Leonard Bradley, a descendant of a Ball family in Connecticut "that the word Ball means bold, courageous, and implies that the first owner of the name showed himself to be possessed of these qualities in a remarkable degree."[7]

BY CONTRAST, THE WASHINGTON NAME, WHICH CAN BE TRACED TO A TOWN in northeast England from the late twelfth century, has a well-documented history. "The Washington family is of an ancient English stock," able to be traced to the 1100s, wrote historian and "The Legend of Sleepy Hollow"–famed author Washington Irving.[8] Though variations of the name "Washington" go back centuries, the lineage can be traced relatively smoothly. Even before George, the Washington name was noteworthy both abroad and in the colonies. This added a certain sense of almost divinity and royalty to the man. "Many of the most illustrious benefactors of mankind have been not less remarkable for the obscurity of their origin than for the greatness of their destiny; but Washington sprung from a family whose name had already become known to history."[9]

In 1183, within the Boldon Book, a survey of landholdings as ordered by the bishop of Durham and so named due to the parish Boldon being listed first, was mention of William de Hertburn, who received a manor in Wessyngton. It read, in Latin,

Begins the book which is called the Boldon Book, in the year
of the Lord's Incarnation 1183 . . . William de Hertburn has
Wessyngton (except the church and the land around the church),
for exchange for the village of Hertburn.

It is possible that Hertburn was from Norman French stock, thus
putting a further possible connection between the Balls of medieval England and the Washingtons and giving a foreign edge
to a purely American figure. From there, William de Hertburn
took the name of his manor, Wessyngton, and became known as
William Wessyngton. His family lived in modern-day Washington Village, in what's called "Washington Old Hall," which was
rebuilt in the seventeenth century and today on occasion flies the
Stars and Stripes.

Successive generations changed the town of Wessyngton to
Weshington in the middle of the fourteenth century, and again to
Wasshington and then eventually to the now-familiar Washington. Irving believed that the town of Wessyngton and its varia-
tion of Wassengtone "is probably of Saxon origin," and was cited
in Anglo-Saxon chronicles in the tenth century. Another source
speculated its etymology came from the Old English "hwaes,"
meaning "sharp" or "keen."

A sensational headline in an article written in the *New York*
Times from 1911 by Professor Bernard Cigrand of the University
of Illinois screamed, "WASHINGTON NOT REAL NAME OF
OUR FIRST PRESIDENT." Factually correct and unnecessar-
ily dramatic, the article continued to say that the descendants of
the Wessyngton/Washington manor persevered through Scottish
raids and attacks, demonstrating the family's loyalty to the English Crown and kingdom.[10]

NOTABLE BEARERS OF THE SURNAME "BALL" TENDED TO LIVE UP TO ITS ET-
ymological meaning: "bold." For instance, the priest John Ball,
who led the anti-establishment and reform revolt of peasants in
1381, was a Lollard—a sort of proto-Protestant. He deemphasized

hierarchical Christianity as seen in the ecclesiastical system and the ritualistic liturgy in the Church. His views were summarized in a famous saying, "When Adam delved and Eve span, Who was then the gentleman?" When earth was populated with but two humans, they were equal. There was no king, no nobility, no priestly class, and no popes or cardinals or bishops. Just two humans, on equal footing. It was provocative and insulting for both the Church and the Crown.

The Peasants' Revolt against the fourteen-year-old King Richard II ultimately failed and Ball, called "the mad priest of Kent," was hanged, drawn, and quartered on July 15, 1381, at Saint Albans Abbey.

Other Balls through the centuries and through the English country existed, some who may or may not have been directly related to the Balls of Virginia. These included Francis Ball, an early settler of Springfield, Massachusetts; Frances Ball, a Dublin nun; and many others, some of whom came from eastern England, and some of whom came from the western part of the country.[11]

But bold they all were. "Possibly," speculated prominent nineteenth-century historian Benson Lossing, "the democratic spirit of our beloved patriot was inherited through a long line of ancestry . . ."[12]

THE BALL FAMILY OF VIRGINIA CAN BE TRACED MORE CONCRETELY AND with much less guesswork to the sixteenth century, with William Ball, the great-grandfather of George Washington and Mary Ball's grandfather. This William emigrated from England to the colonies, and, depending who is asked, was the son of two fathers. Some say his father was the Reverend Richard Ball of Northamptonshire, England, vicar of Saint Helen's in Bishopsgate, London,[13] while others, including Mount Vernon's official website,[14] believe it was William Ball of Lincoln's Inn, London.[15]

WHETHER THE NEWLY ARRIVED VIRGINIAN WAS WILLIAM SON OF RICHARD or William son of William, the great-grandfather of George

Washington was a man worthy of his descendants. He was born around 1615, and "practically nothing is known of his early life, and we can only infer that he was educated in or about London," said Heck.[16] London was the cultural center of England, and either Reverend Richard or William Ball would have educated his son in the vicinity.

On July 2, 1638, at around twenty-five years of age, William married Hannah Atherall, who came from a London family and whose father was a "barrister-at-law," a lawyer who specialized in English common law.[17]

THE MARRIED COUPLE WILLIAM AND HANNAH BALL HAD FOUR CHILDREN— three sons, one daughter—in England: the eldest, Richard; William, born in 1641, and often distinguished by his father as William Jr. or "Captain William"; Joseph, born 1649, the father of Mary and grandfather of George Washington; and Hannah, born 1650.[18]

William Ball, about forty years of age, emigrated to Virginia around 1657, fleeing England during the English Civil War. The conflict was principally fought between those who supported more power to Parliament (dubbed the Roundheads), and those who believed in the royal prerogative (the Cavaliers), in support of the king, then Charles I. The Roundheads were mostly Puritans from southern and eastern England, made up mostly of gentry and merchant classes; whereas the royalists and Cavaliers were nobles and Anglicans and Catholics from north and western England.

A turning point in the war was the Battle of Marston Moor in 1644, where William Ball participated as a staunch royalist and served "faithfully under the banners of the ill-fated Charles." Heck continued on the topic of Marston Moor: "By the defeat of the Royal Army, Colonel Ball lost the greater part of his estates, which were by no means inconsiderable."[19] The battle resulted in a major defeat for the royalists at the hands of Oliver Cromwell, and through a series of captures, defeats, and recaptures, Charles I was convicted and beheaded for high treason in 1649.

A royalist like Ball had to go, lest he meet the same fate, so he went to the Colony of Virginia, "the most loyal of the king's possessions," setting foot in the New World around 1650.[20] A comprehensive list of early Virginian immigrants compiled by George Greer, a clerk at the Virginia State Land Office, at the turn of the twentieth century, noted that there were at least twelve other Balls that came to Virginia between 1623 and 1666, with a "Ball, Wm" arriving in 1653.[21]

If relations to Preacher John Ball encouraged George Washington in his desire to break away from authority, a later descendent, George Washington Ball, took a much more pessimistic view: "Poor *'Bals'*! Rebellion seems to have run in their blood, and their ill-luck to have led them generally on the losing side."[22]

WILLIAM BALL'S REASONS TO FLEE WERE NOT UNIQUE TO THE ENGLISH people. Of the many incentives that immigrants to the New World had to flee the Mother Country, the desire for freedom was one of the more passionate. Quakers like George Fox arrived from England to get away from institutional persecution. Puritans came to New England, Catholics to Maryland—all of different faiths in many different ways, running west to their own freedom.

Political persecution, of course, was a fear both large and small. Sir William Berkeley, to be royal governor of Virginia, encouraged royalists to go to Virginia in hopes of keeping a royal colony. However, it was overestimated: "Cavalier immigration to Virginia during the 1640s and 1650s was not large; no more than a couple of hundred arrived in this period. The great majority remained in England either in retirement or fomenting plots against Parliament and the republic. Others fled to the Continent."[23]

WILLIAM BALL LEFT HIS NATIVE LONDON AT A TIME OF UPHEAVAL, NOT just politically but environmentally. London was a plague-filled city of smog, filth, and pollution. Among those blamed for the air quality were "Brewers, Diers, Sope and Salt-boylers, Lime-burners, and the like," according to John Evelyn, a member of the newly founded Royal Society of London for Improving Natural Knowledge. Therefore, he proposed to move all of these industries from the city, approximately six miles away. Additionally, planting flowers and other greenery, like lime trees, yellow or white jessamine, or lavender throughout the city would help dissipate some of this "smoak," and from there London could be seen restored. "Men would even be found," he said, "to breath[e] a new life as it were, as well as London appear a new city, delivered from that . . . so infamous an aer."[24]

The New World's rural and open land, away from the pollution of Europe, would offer a clean breath of fresh air for all immigrants, especially William Ball, who needed a fresh start desperately.

IT WAS ONLY THREE GENERATIONS FROM WILLIAM TO GEORGE, AND A LIT-tle more than a century from native Englishmen loyal to the Crown to Americans fighting against it. It is with a bit of irony, then, that genealogist Earl Heck believed "he and his family lived with relatives waiting for the Stuarts to be restored to power in England."[25]

When the English Crown was restored and the republican cause squashed in England in 1660, William, the royalist, did not return. He had become fully Virginian. "He soon decided to cast his lot with the fortunes of the New World. After 1660, he was a member of a court to make a treaty with the Indians and to establish a boundary for the occupation of land by the white man."[26]

His first occupation, according to Northumberland County records, was "Merchant," which suggested that he did not have a plantation or land of his own. It wasn't until 1663 that he became a "planter" by occupation, after he had a grant of land issued on January 18, 1663, on Narrow Neck Creek in Lancaster County.

It was three hundred acres of land, and previously belonged to neighbor David Fox for ten years.[27] This would later be called Millenbeck, fifty acres of which, within several dozen years, was bought by the County of Lancaster to create the short-lived but nevertheless thriving county seat of Queenstown.

Within a decade, William's children, wife, servants, and slaves also emigrated to the Colony of Virginia. He brought his son Richard on May 27, 1657, and his son Joseph in early January of 1660. On March 4, 1665, his wife; his son William; his daughter, Hannah; and others arrived.[28]

Within a decade, the Ball family's move to the New World was complete.

LANCASTER COUNTY, VIRGINIA, WAS FRONTIER LAND. MANY PEOPLE wanted to lay claim to the rich, fertile region, the hostile Native Americans notwithstanding. The Virginia Assembly attempted but failed to legislate who could travel into the area, and soon many took up acreage. In 1643, Captain Samuel Matthews acquired four thousand acres of land north of the Rappahannock River. Sometimes this acreage was taken by diplomatic trade and purchase from the Native Americans, such as in the late 1650s, when Colonel Moore Fauntleroy bought land from the Rappahannocks. Again, legislation was passed to curb and protect the ever-fickle relationship between colonizers and natives. "Even so," wrote historian James Horn, "English settlers appear to have taken the view that, once they moved into an area, resident Indians would soon move out." This proved true for many tribes.[29] By 1653, ninety-one families had settled along forty-five miles of the river. Through the years and decades following, the vast majority of those settling in Lancaster were immigrants from England, from all backgrounds. There were some Scottish, Welsh, and Irish immigrants, but those were few and far between. Of course, indentured servants also provided a major population boost. By 1656, they accounted for almost half the population of Lancaster County.[30]

HISTORIANS, IN ORDER TO DISTINGUISH BETWEEN WILLIAM BALL THE IM-
migrant and his son William (not to mention this second William's
descendants named William), use the moniker "Colonel William
Ball" for the former. He received that rank in or around 1672,
and Horace Hayden believes it was as he was appointed county
lieutenant of Lancaster. "Military titles were never assumed in
those days," he wrote. "They were conferred by the authority of
the Governor, who, under the Royal Charter, was Commander-
in-Chief of the Colony. . . . Such was the danger to the colonists
from the incursions of the neighboring Indians that some show of
military organization was necessary for defense." To wit, Colonel
Ball's son William was also named captain, according to transac-
tional records.[31] He mobilized militia and men, led under Nathaniel Bacon, for Lancaster County against Native Americans, and he
was to "make choise of the men and horse before lymitted in their
countyes to be raised for their respective fforts."[32] His responsibilities in Virginia more than likely made him come in contact with
the Washington family as well.

Nevertheless, this Colonel William Ball gained the trust of the
New World settlers, witnessing deeds almost immediately, before
his children or wife arrived. He was clearly an active member
of the community. By 1653, he witnessed the patent given from
Henry Fleet to John Sharpe, two hundred acres of land in "Fleet's
Bay," Lancaster County.[33] He was granted passage at least four
times in intervening years to England. By 1659, he was appointed
Commissioner of Lancaster County, witnessed more wills, and inventoried and appraised several estates as necessary.[34] He even
handled settlers' legal disputes, suggesting that he was a prac-
titioner of law.[35] He would continue handling such disputes
throughout his life. By 1670, he became Chief Magistrate of Lancaster, a high position worthy of the Ball name, working under
the Virginia legislature based in Jamestown.[36]

In 1667, the Colonel became Major William Ball, and on September 30 of that same year, he received "240 [acres] of land on
the N. side Rapp'k adjoining the land of David Fox."[37] The land of

Major Ball started to pile up, and by the time of his death in late 1680 he had accumulated "nearly two thousand acres of land in Lancaster and Rappahannock Counties," as counted by Heck.[38]

THE PLANTATION HOME WILLIAM OWNED IS LONG LOST AND GONE, MAKING it difficult to say where exactly this family lived. Wherever it was, it was built by carpenter and wheelwright Edward Floyd, a specialist in the trade in Lancaster. He was described as building "the great home" of William Ball, and he later repaired and did extensive work on the windows, chimneys, and rooms of William's son's "great dwelling house."[39]

WILLIAM BALL THE IMMIGRANT DIED IN LATE 1680, AROUND THE AGE OF sixty-five, survived by his wife, Hannah, and three of his four children. His will, dated in October of that year and sealed by Thomas Everest and John Mottby in November, reads as a man who came from simple and humble origins, relatively, amassing quite a hoard of wealth within thirty years from emigration to the time of his own death.[40]

Of particular note in the will was the very little that he gave his daughter, Hannah: "Only five shillings." All his other possessions, including land, cattle, and slaves, went to his wife or his other children. Hannah Ball got married; as Hannah Ball Fox, she had her own life and was no longer part of the Ball family.

William's wife, Hannah, made out her will in late 1694, almost fifteen years after her husband's death; however, her son William, their oldest surviving child (Richard had died sometime earlier), died in November of that year, making much of the will irrelevant. On June 25, 1695, Hannah wrote her second will—just in time, too, as she died shortly after. She had not remarried, a custom unusual for the time, instead continuing her late husband's work in quiet solitude.[41]

Both William and Hannah, the grandparents of Mary and great-grandparents of George Washington, were buried in unmarked graves—or, at least, by the time of Joseph II's interest in his family's whereabouts a century later, they were unmarked. In several letters to cousin Joseph Chinn between 1754 and 1755, he requested, "I would have you out of hand take a good hand or two with you, and go down to the Plantation where my Grandfather and mother lived and are buryed; and get the assistance of Hannah Dennis to shew you as nigh as she can the spot where they are buryed; and let the hands skim the ground over about four or five inches deep; and if you come over the graves you will find the ground of a different Colour. If you can find that, then stake it out at the four Corners with sound Locust or Cedar stakes, that you may be sure to find it again; for I think to send a stone to put over them. Pray take diligent Care of this affair, and you will oblige."[42] After one final request to "dig a little deeper" in late 1756, the matter was dropped and the graves forgotten.[43]

COLONEL WILLIAM BALL'S YOUNGEST SON WAS JOSEPH, BORN ON MAY 25, 1649, in England, who would later become Mary's father. The other branches of the Ball family continued in Millenbeck, Virginia, and all around the Colony of Virginia for the next three centuries.

There was a catch, however: "Of [Joseph] very little is known," wrote genealogist Horace Hayden.[44] He married twice, first around 1675 to Elizabeth Romney (or Elizabeth Rogers).[45] Like his father, he was prominent in the Lancaster community. Joseph became a vestryman and worked closely with the local Anglican church, Saint Mary's Whitechapel in Lancaster, built several years earlier in the late 1660s. The Balls were closely knit to this church, placed squarely on the main county road, with relatives both distant and close having worshiped in this denomination for centuries. To this day the cemetery holds the remains of many of the Virginia Balls, known and unknown, famous and not, unsung heroes of this family. In the 1740s, when the church was expanded and rebuilt, two Ball relatives were tasked with and financed a south gallery in the church, which is noted prominently on Whitechapel's website and in many publications as a source of pride of its rich history.[46]

Around 1700, Elizabeth died, and Joseph remarried a woman of unknown background named Mary Johnson around 1707. This Mary was soon after to give birth to their only child, a daughter named Mary, who became the mother of George Washington.[47]

WHAT ABOUT THE REST OF THE BALL FAMILY?

Much like the rest of the early Balls, and of Mary the mother of Washington herself, the truth is a mixture of legends and facts and records and oral traditions. "History is silent" about George's maternal grandmother.[48]

"The maiden name of Mary Johnson is unknown. Some researchers believe it was a Montague, because her grandson, George Washington, used a flying griffin similar to the one on the Montague crest as his personal seal. Others think she was a Bennett."[49] Hayden did give credence to the idea of a Montague lineage, making George related to Drogo, who nearly seven centuries earlier invaded England with William the Conqueror.[50]

The Montague connection was shrugged off by some biographers, most notably Sara Pryor (who had a penchant for placing emphasis on the romantic rather than the factual side of history), as nothing more than a historical anecdote worthy of a mention. "It matters little whether or [not] the mother of Washington came of noble English blood," she wrote. "For while an honorable ancestry is a gift of the gods, and should be regarded as such by those who possess it, an honorable ancestry is not merely a titled ancestry. Descent from nobles may be interesting, but it can only be honorable when the strawberry leaves have crowned a wise head and an ermine warmed a true heart."[51]

The other side of the spectrum as to who Mary Johnson was is quite anticlimactic. The mother of Mary Ball Washington, some report and have concluded, was simply and boringly Joseph's housekeeper during his first marriage, based only on her witnessing a deed some years earlier. Other genealogists and those interested in Ball history have balked at the idea; Hayden, for instance, immediately scolded those who debased Mary Johnson as nothing more than "plebeian,"[52] a rumor made more egregious from the supposed gentry origin of the family. Yet the rumor continued on; James Flexner in his multivolume biography of George Washington perpetuated this legend, calling Mary "an illiterate widow."[53] And from there the appellation stuck.

This controversy serves to illustrate an important point. Despite Americans' new democratic ideals, they still found a need to seek a mythological origin story for this country of shoemakers and silversmiths. The allure of the divine right of kings lingered in the air.

FROM AROUND THE TIME OF HIS MARRIAGE TO ELIZABETH IN 1675 TO HER death a quarter century later, Joseph and his family returned to England, and it was there that "his children by his first marriage were born."[54] Yet his return to Virginia and building of an estate at "Epping Forest" allowed his second wife, Mary, and Virginian-born daughter, Mary, to enjoy a purely American lifestyle.

The location of their estate on the 720-acre land near the Rappahannock was originally called "The Forest Plantation" or "The Forest Quarter" or simply "The Forest," only being called "Epping Forest" in the nineteenth century by a Ball descendant. It was situated in the upper part of the county. Near Epping Forest in Virginia was Bewdley, the plantation home of James Ball, third son of Captain William Ball, son of William, built about 1700. Before it burned down in a fire in 1917, the home housed several generations of this Ball branch. Rumor had it that the house, which overlooked the Rappahannock, provided well-needed signal lanterns at night, through its many windows, for passing Continental and American ships during the Revolutionary War and the War of 1812.[55]

The name Epping Forest was most likely a reference to their ancestral lands. In England, stretching for eighteen miles and containing nearly six thousand acres from northern London into Essex, there still stands an ancient forest, once owned by royalty, scattered with manor houses dating from the medieval period.

Legend has it that Boudica, the famed female warrior, made her last stand there two millennia ago. This was the old and wondrous Epping Forest. The forest's possession stretches back to the Iron Age, occupied by the Romans, by commoners, by royals and nobles and abbeys. (It is currently owned and managed by the City of London.) In Edward North Buxton's guide to Epping Forest in 1905, he summarized, "From pre-Norman times until the eighteenth century, the maintenance of the sporting rights of the sovereign was paramount, and to this every other interest was subordinate." Only the king could hunt there, while commoners were permitted only to graze or raise animals and gather wood.[56]

This same area, near the village of Epping, may be the ancestral land of the Ball family. Joseph Ball, the older half brother of Mary Ball Washington, the son of Joseph her father, lived in Stratford, in Greater London, in the mid-1700s, expanding the connection to the area.

ABOUT TWO YEARS BEFORE THE MOTHER OF THE FATHER OF HIS COUNTRY was born, a different Founding Father opened his eyes for the first time in Boston, in the colony at Massachusetts Bay. His name was Benjamin, born to Josiah and Abiah Franklin. Decades later would see Benjamin and a decades-younger George Washington working side by side, along with others of all ages, in support of an entirely different new world than that which they or anyone had been born into.

THE BALLS, RIGHT FOR THEIR ETYMOLOGICAL ORIGINS, WERE BOLD. Whether it was John Ball the preacher; or the Balls who may have come from Normandy; or Colonel William Ball, the first in Virginia, all took a plunge for the betterment of their family and family name. It was in character for the Balls to flee war-torn England after the defeat of the royalists to seek a new life in Virginia. It took less bravery to stay than to flee, lest William Ball find himself beneath the blade of an executioner, just like King Charles.

The motto of the family coat, "Caelumque Tueri," was taken from the last two words of this poem by the ancient Roman poet Ovid:

Pronaque cum spectant animalia caetera terram
Os homini sublime dedit caelumque tueri.[57]

"Whereas other animals observe to the earth, he gave the face of man sublimity and to look to the sky."

"To look to the sky"—their motto resonated not just for the Balls, but eventually for all Americans. The Balls looked to the horizon toward the New World and to Virginia—and set their course to become more than even they would have imagined.

STUDIES OF THE WASHINGTON NAME RANGE FROM THE FANTASTICAL TO the mundane. This was a dialectic in the nineteenth century as historians wished to study the founder of our nation. The United States has no king, so make George descended from royalty, wrote some. The United States has no nobility, so emphasize his ancestors' noble titles, wrote others. The United States has no official religion, so emphasize his saintly ancestry, reflected a third group. Simultaneously, American citizens wanted humble origins reflecting the American dream . . . contradicting the desire to place George in the pantheon of nobility of the Enlightenment.

Much like that of the Balls, the Washington lineage was riddled with sons named after their fathers, or brothers named after uncles. What the Balls had with their cross-generational Williams and Jameses, the Washington men had with Lawrences and Johns. George Washington's grandfather was Lawrence Washington; Lawrence's father, the first of the Washingtons to settle in the Americas, was John; John's father was Lawrence, a reverend of Purleigh, Essex; the Reverend Lawrence's father was Lawrence, who had a father named Robert, whose brother was Lawrence; Robert's father, the builder of Sulgrave Manor in central

England, was Lawrence. And Washington's older stepbrother was Lawrence.

Some historians, studying the paternal ancestry of one of the greater Founding Fathers, desired to connect his bloodline to literal divinity.

Albert Welles, in his 1879 book, believed George Washington could be traced to the Norse god Odin, or at least his pseudo-mythic and -historical equivalent, giving it the less than subtle title *The Pedigree and History of the Washington Family: Derived from Odin, the Founder of Scandinavia. B.C. 70, Involving a Period of Eighteen Centuries, and Including Fifty-Five Generations, Down to General George Washington, First President of the United States.* Welles was no conspiracy theorist or crackpot historian; he was the president of the American College for Genealogical Registry and Heraldry.

Welles, in the preface, cited an anonymous source from London who had "been engaged for thirty years in gathering evidence" on this. This source believed that without his research "that great man's lineage would not have been revealed."[58] The book analyzed generation after generation of Washington ancestry, and concluded that he was the ancestor of Odin, who around 70 B.C. was the founder of Scandinavia and "the son of Fridulf, supreme ruler of the Scythians, in Asaland, or Asaheim, Turkestan, between the Euxine and Caspian Seas, in Asia. He reigned at Asgard, whence he removed in the year B.C. 70, and became the first King of Scandinavia. He died in the year B.C. 50."[59] Ultimately, the author traced Denmark royalty from father to son to the American rebel.

There's cult of personality, and then there's this. The United States did not have a rich genealogical history by 1879 (in fact, it had just undergone a civil war that threatened its foundation's very existence), and presumably many, Welles included, wished to fill that void. There were no mythic figures or kings or queens or empires of the century-old nation on the other side of the Atlantic, so why not find ways to create one?

MID- TO LATE 1500S ENGLAND WAS A GOLDEN AGE FOR THE ISLAND. THE Washingtons' Sulgrave Manor was central not only geographically but also culturally. Not far from the Washingtons was William Shakespeare. According to Ethel Armes, there was an additional connection to American history here. "The ancestors of Benjamin Franklin dwelt for centuries in Ecton, a village about twenty miles from Sulgrave. . . . Four American presidents trace their ancestry to this locality." This includes Washington, John Adams, John Quincy Adams, and the ancestors of Warren G. Harding. It is possible ("more than likely," says Armes, "and how pleasant to fancy it true!") that these ancestors, as well as Shakespeare, met often.[60]

Outside stood a garden, called the Orchard, in which some think Lawrence, possibly planting over the monastery's previous gardens, took inspiration from contemporary stylists, such as Andrew Borde's tome, which stated, "It is a commodious and pleasant thing in a mansyon to have an orcharde of sundrye fruytes."[61] It was a mansion worthy of his name and ancestry.

TODAY, THE SULGRAVE WEBSITE BEARS THE COAT OF ARMS OF THE WASHINGTON clan, which not by coincidence is the same coat as that on the District of Columbia's flag. The Manor is open to the public during certain hours of the day.

Upon the dedication and opening of the Manor in 1914, the *New York Times* ran a full-page story with pictures. "A charming piece of old architecture," it read, "gray with the rains, frost, and sunshine of 300 years."

At the end, the article prophetically stated, "In dedicating the Manor as a memorial to the peaceful relations existing between the two great English-speaking nations during a century, the British Committee has created a permanent memorial of permanent interest."[62]

LAWRENCE WASHINGTON, THE BUILDER OF SULGRAVE MANOR AND AN IMportant man in his own right, died in February of 1584, at over the age of eighty. The anti-Catholic and tumultuous times of King Henry VIII and the short bloody years of Queen Mary were long over, and Queen Elizabeth I was fewer than thirty years into a forty-five-year reign.

It was the Elizabethan Age.

THE NEXT SEVERAL GENERATIONS OF WASHINGTONS CONTINUED SIMILARLY; they neither started revolutions nor built exuberant manors, nor did they supposedly hide kings and queens from persecution.

Lawrence Washington, the grandson of the Sulgrave builder, was born there around 1568. He had married a Butler, Margaret, whose family descended from the medieval Plantagenet dynasty. As a descendant of King Edward I of England, who reigned in the early fourteenth century, this would place all of her children, and children's children, ad infinitum, as royal blood, George included. Triply so, as Margaret and thus her children and descendants could claim both the king of France and the king of Spain as ancestors, the latter of whom, King Ferdinand III of Castile, the father of Edward's wife, was canonized by the Catholic Church, thus making George "an inheritor of the Saints in light," as "in his veins flowed the blood of the Servants of God," as he was deemed by the monthly *Catholic World* in 1916.[63]

Lawrence Washington, husband of Margaret, died in 1616, after having at least four children.

It was this next generation that brought American interest.

Born around 1602 in Sulgrave Manor, Lawrence, the great-great-grandfather of George, became a lector at the University of Oxford, and later a reverend and rector of Purleigh in Essex; he married and had children. He was a father to a pair of brothers, Lawrence and John, who years later would sail to the colonies.

That's when England changed.

The English Civil War in the mid-1600s brought not only a political change to the kingdom but also a religious change. Puritanism became almost an official religion of the country;

Christmas was banned in 1647 and a strict if not prohibitionist mentality on alcohol spread throughout. Woodrow Wilson, future president, in his biography of George, wrote that Reverend Lawrence "had been cast out of his living at Purleigh in 1643 by order of Parliament, upon the false charge that he was a public tippler, oft drunk."[64] Another source repeats this, that he was "allegedly ousted by the Puritans for drunkenness."[65]

The English Parliament, taking on its Puritan tenor, described him as one of the 2,800 Anglican clergymen who were "scandalous, malignant priests," saying that Lawrence was "a common frequenter of ale-houses, not only himself sitting daily tippling there but also encouraging others in that beastly vice. [He] hath often been drunk."[66]

The truth was, however, that he was simply a royalist, from a royalist family. Yet his career and livelihood were destroyed.

He died only a few years later, in 1652.[67]

What was here in England for royalists and ancient Englishmen like the Washingtons? Certainly any pension would have been spread too thin among a widow and her children. Competing political and religious faith were all but banned in a Puritan do-or-die society. Where did one go?

West.

WHILE WILLIAM BALL WAS SAILING TO THE NEW WORLD, SO TOO WERE THE sons of Reverend Lawrence, including the eldest John, born in 1632, and Lawrence, born in 1635. The same decade that the Balls were fleeing England, the Washingtons were also arriving in Virginia.

They arrived around 1657 or 1658. Two years later, in 1660, in England, the monarchy was restored with the ascension of King Charles II.

But by then, England was behind the Washington and Ball families.

Moving to the New World was not a simple task nor one that

should have been taken lightly. That was true for the Balls, and that was true for the Washingtons: when they "came to people the New World . . . a coat-of-arms had slight bearing beside the qualities of personal force and distinction."[68] It was a new beginning in the colonies.

John Washington, brother of Lawrence, son of Reverend Lawrence, great-great-grandson of Lawrence of Sulgrave Manor, and great-grandfather of George Washington, arrived and settled at Bridges Creek in Westmoreland County, Virginia. "What was poverty in England," wrote Henry Cabot Lodge in the late nineteenth century, "was something much more agreeable in the New World of America."[69]

THE PLANTATION AT BRIDGES CREEK LATER EXPANDED INTO NEARBY POPES Creek, where an estate, later known as Wakefield, would become the birthplace of George Washington himself. The estates on both Popes and Bridges Creek were typical Colonial houses, two stories with a number of rooms and chambers.[70] John met and married Anne Pope, daughter of Nathaniel Pope (whom the creek was named after, and a major founder of the Colony of Maryland[71]), and they had their first child, Lawrence (grandfather of George).

The numbers of his acreage soon started to add up, making a man rich from tobacco even richer. His father-in-law gave 700 acres to the newlywed couple in Mattox Creek, and John purchased, in the spring of 1659, up to an additional thousand acres (700 of which went to his brother Lawrence when he emigrated), and in 1664 he purchased 1,700 acres of land, where the future Wakefield was to be built. Through the years, he purchased several hundred more acres in the county, eventually tallying up to a total of ten thousand.[72] His brother Lawrence, soon after his arrival, settled farther south along the Rappahannock River.[73]

JOHN WASHINGTON WAS ELECTED AS A MEMBER OF THE VIRGINIA HOUSE of Burgesses and raised to the rank of colonel in the militia. In

1664, the parish he attended was renamed Washington Parish in his honor, and he was enlisted among other commissioners in March of 1675 or 1676 "to use Indians in the warre and require and receive hostages from them, alsoe to provide one hundred yards of tradeing cloath to each respective ffort, that it be ready to reward the service of Indians."[74] In this war the Native Americans gave him the nickname "Conotocaurious," meaning "devourer of villages," and he often assisted Marylanders in attacks on and defense from Native Americans, at one point killing six chiefs.[75]

BUT THERE WAS ANOTHER WAR ITCHING TO BE FOUGHT.

Nathaniel Bacon, a prominent Virginian settler, rose in 1676 against the governor, William Berkeley. Bacon was disillusioned by Virginia's lack of interest in the constant Indian attacks on the frontiers, which were increasing, and the counterattacks by the colonists. He raised his own militia without first consulting the governor. Berkeley declared him a rebel, and Bacon, in retaliation, marched into and burned down Jamestown.

John Washington joined one of the two sides. He could have joined Berkeley, alongside William Ball himself. "The Fates that move the pieces on the chess-board of life ordained that two prophetic names should appear together to suppress the first rebellion against the English government," wrote Sara Pryor in her typical poetic prose.[76]

On the other hand, different sources, including Joseph Sawyer in his 1927 biography of George Washington, clearly stated that John "joined Nathaniel Bacon—often called 'the young Cromwell'—in hurling defiance at loot-saturated Governor Berkeley of hated memory."[77]

This could foreshadow the Washington descendant's penchant for revolutionary fervor.

COLONEL JOHN WASHINGTON, THE GREAT-GRANDFATHER OF GEORGE, DIED in 1677 at around the age of forty-five, having amassed, in the end, 8,500 acres of land and having made an undeniable mark on Virginia's history.

His eldest son, Lawrence, to become the grandfather of George, received 1,850 acres.[78] Total acreage was worth approximately $70,000.[79]

John Washington's will stated, "I give unto my sonn Lawrence washington my halfe & share of five thousand acres of Land in Stafford County which is betwixt Coll Nicolas spencer & myselfe which we [are engaged]."[80] That five-thousand-acre land was called Little Hunting Creek Plantation, the site of what would later become known as Mount Vernon.

LAWRENCE, GRANDFATHER OF GEORGE AND ELDEST SON OF COLONEL JOHN, was born around 1659 in Virginia, becoming the first of the Washington clan to be born in the colonies, possibly at Bridges Creek.

Lawrence served until his death in 1698 as a Virginian justice of the peace and traveled to England at least once in the spring of 1686 for unknown reasons. In 1695 he served as high sheriff of Westmoreland County. He had several slaves, as indicated by court records.[81]

Around the age of twenty-seven, John married Mildred, the daughter of Colonel Augustine Warner, of another prominent family. They had three children: John, Augustine, and Mildred. The oldest, John, was eight years old when his father died. Mildred, his daughter, was an infant. Augustine, the future father of George Washington, was only four years old when he became fatherless.

To his daughter, the older John gave Little Hunting Creek Plantation, of 2,500 acres; to John, his eldest, he gave his current residence of Popes Creek; and to Augustine he gave 1,100 acres. In accordance with his will, if his children were to be underage when he died, that they would "continue under the care & Tuition of their Mother till they come of age or day of marriage, and she to have the profits of their estates, toward the bringing of them up and keeping them at school."[82]

FROM ENGLAND TO VIRGINIA, FROM RICH TO POOR TO RICH, THE WASHING-
tons endured war, oppression, and bounty. It would not end.

WITHIN A FEW SHORT DECADES FROM THE DEATH OF HIS OWN FATHER, AU-
gustine Washington married Mary Ball, of an equally rich and
equally early Virginian family.

Chapter 3

The Rose of Epping Forest

EARLY LIFE OF MARY BALL

1708–1730

"I am now learning pretty fast. . . ."

B elle of the Northern Neck," "Rose of Epping Forest," "Toast of the Gallants of her Day": these were some of the sobriquets passed down in biographies and hagiographies of Mary Ball. All were romantic, and all were probably simple hearsay and nicknames coined decades after her death. But whether these nicknames were created during, after, or far after her long life in Virginia, they all gave a sense of beauty, attractiveness, and purity of the young Mary, adding a mythical aura.

The early life and childhood of Mary Ball is shrouded in mystery and a simple lack of facts. Like the genealogy of her ancestors, and like the genealogy of her future husband's ancestors, all that happened was simply not recorded. And what was recorded did not necessarily happen. By 1850, only a couple of generations after Mary's passing, Margaret Conkling, one historian in the mid-nineteenth century, noted that no record of her childhood was preserved; in successive years, some authors, including famed historian Benson Lossing, saw fit to either romanticize or simply assume rumors and tales and forged documents as facts.

MARY WAS BORN AROUND 1707 OR 1708 TO JOSEPH AND MARY JOHNSON BALL, the only child of their marriage. Genealogist Horace Hayden argued that "she was born as late as August 25, 1708, New Style," on account of her age at death eighty-two years later.[1] Another source, author Nancy Turner from the early twentieth century, speculated early November 1707, on an autumn day.[2] And yet a third source decades later stated she was "apparently" born during the winter of either 1708 or 1709.[3]

Though the exact date or year is unknown, a fight between historians had taken place, with the implications of some years over others. Benson Lossing, whose historical works were gospel to the nineteenth-century American, said flat out that Mary was born in late 1706.[4] Mary Terhune, writing as Marion Harland in her *Story of Mary Washington*, repeated similarly that it was an "autumnal day in the year of our Lord 1706."[5] From there the year 1706 was repeated continuously, as in an article in *The Spirit of '76*, among many others.[6] Though poetic that the mother of Washington was born exactly seventy years before the year of independence, there was a serious, and possibly damning, implication: if Mary was born in 1706, as attested, then that would mean she was born out of wedlock. Hayden pointed to a deed between her father, Joseph, and his son on the seventh of February, 1707, in which the former stated that "at this date I have no wife," so any date before that directly accused her of illegitimacy. Laws were in place for bastard children, resulting in fines and additional punishment to the parents, *especially* if master and servant were involved. Mary Johnson, the mother, was rumored without foundation by several sources to have been the maidservant of the Balls. A "freeman" and maidservant having an illegitimate child would have resulted, per an act of the Virginia General Assembly in 1657 to 1658, in paying "five hundred pounds of tobacco to the vse of the parish where the said act is comitted or be whipt."[7] That's not exactly a light sentence by any means. For this egregious assertion, Hayden called Lossing's book "historical fiction," further gnawing back, "I do not wonder when Lossing, an historian, has set his pace by his romances."[8] In another publication, his gnaw became a bite: these anecdotes that were repeated time and time again "destroy faith in historical writers,"[9] he said.

All is fair in love and war . . . and historical research, apparently.

Whether or not her birth was in 1706 or 1707 or 1708, and whether or not Mary Ball was born in or out of wedlock in the Northern Neck, it "came without fuss and feathers . . . Colonial families increased with such rabbit-like regularity that there was not much more excitement over a lying-in than over the making of an apple pie. . . . [It] was merely an episode in the daily life at Epping Forest."[10] A lack of extant records may prove this, and multiple children through previous marriages may have diluted the exceptional circumstances of a new life.

WHERE WAS SHE BORN? THAT LOCATION WAS MOST LIKELY EPPING FOREST Plantation in Lancaster County, owned by her father. This building currently sits less than a mile from State Route 3—which is also appropriately known as Mary Ball Road, noted for its pass through historical Virginia and for having possibly been a major colonial route and winding through Lancaster County, through the small town of Lively, moving northward through farms and forests and Balls Branch, until it merges with River Road to become History Land Highway right at Chinns Pond. Moving southward from Lively, Mary Ball Road hits Lancaster, where the Mary Ball Washington Museum and Library stands, down about seven miles to Kilmarnock. This fifteen-mile stretch of road commemorates this event on this unknown day in this unknown year, with signs pointing to Epping Forest.

Early-twentieth-century biographer Nancy Byrd Turner succinctly described the period of Mary's birth in the early eighteenth century as "peaceful air."[11] Mary was born as Virginia was relatively settled and calm, the threats of Indian attacks or counterattacks minimal. England's civil war half a century earlier was over, and the monarchy had been restored from the hands of parliamentarians.

An unimportant birth to her parents, maybe, during an unimportant time—but certainly not for the United States.

BABY MARY WOULD HAVE SLEPT IN A CRADLE OF WOOD OR WICKER, COVERED with quilted blankets, homemade. It was custom in the colonies to have the initials of the child embroidered on the blanket. Her clothes as an infant were made from linen, most likely, according to Alice Earle, whose book *Child Life in Colonial Days*, while focusing on New England, did allow some thoughts on the Virginian lifestyle. "Linen formed the chilling substructure of their dress, thin linen, low-necked, short-sleeved shirts; and linen even formed the underwear of infants until the middle of this century. These little linen shirts are daintier than the warmest silk or fine woollen underwear that have succeeded them; they are edged with fine narrow thread lace, hemstitched with tiny rows of stitches, and sometimes embroidered by hand."[12]

MARY HAD NO BROTHERS, SISTERS, OR COUSINS AROUND HER AGE. THE youngest child of her father's first marriage, Mary's half brother Joseph Jr., went to England, married Frances Ravenscroft, and settled in his ancestor's country as a lawyer in London by the time of her birth.[13]

Mary was practically brotherless and sisterless living in Epping Forest with her mother and father. Within a few short years, she was also fatherless.

The "'Rose of Epping Forest' was a tiny bud indeed when her father died. . . ."[14]

The first evidence of Mary Ball's existence did not come from her birth or baptismal records, which, if even recorded, have long been lost, but from the death of her father, Joseph Ball, who was around fifty years old at that time. His will, made out on June 25, 1711, and recorded sixteen days later on July 11, was a lengthy and dry document. Joseph Ball died shortly thereafter (one historian on Mary Ball mistakes this will for his father's, saying that he was "lying upon the bed in [his] lodging chamber"[15]). To his youngest child, Mary, about three years of age, he mostly gave:

> *Item I give and bequeath to my daughter* Mary *four hundred acres of Land Lying in Rchmd County in ye freshes of Rapphn River being part of a pattin for sixteen hundred acres of Land to her[.] The s[ai]d* Mary *and the heirs of her body Lawfully to be begotten for ever* . . .
>
> *Item I give to my daughter* Mary *my negro boy Tome and ye negro Joe and Jack yt formerly were belonging to* Jo Carnegie, decd.
>
> *Item I give to my daughter* Mary *all my feathers yt are in ye Kitchen Loft to put into a bed for her.*

Joseph gave most of his estate and possessions to his older son Joseph, Mary's half brother. To his wife, including slaves, "one half of all ye posessions and corn [that] is now ye house for her better support and maintenance," effectively setting her up for life.[16] At least a third of the estate was required by law to be given to a widow.

He died in July of that year, less than a month after his will was made. An inventory of his estate dated July 25 exists, suggesting he died sometime prior. According to the lengthy appraisal, Joseph had clothes and bottles and cases and punch bowls and pots and carpets and table drawers. He had a wealth of possessions. He had cattle of many ages, oxen. He had tables and chests. His estate was valued at well over 350 pounds.[17] He was rich, certainly, though not the richest.

WIDOW MARY JOHNSON BALL REMARRIED, WITHIN A YEAR, EXCHANGING vows with Richard Hues (or Hewes, or Hughes, depending on the document), "a vestryman of St. Stephen Parish living on 161 acres at Cherry Point on the Potomac River side of the Northern Neck."[18]

Cherry Point, Mary's home after her mother's second marriage, was situated between the Yeocomico and Coan Rivers, near the Potomac. The trip from Epping Forest to Cherry Point took the three-year-old girl about fifteen miles through the beautiful Northern Neck of Virginia. For ten years, from three to thirteen, ages bookended with the deaths of her father and her mother, respectively, Mary lived and was raised here.

THE MARRIAGE TO HUES DID NOT LAST LONG. HE DIED IN FEBRUARY 1714, less than two years after his marriage, with no children of their own, when the young Mary was about five years old. She was not mentioned in Richard's will.[19]

First a father and now a stepfather, gone. All before the age of awareness.

FOR THE NEXT SIX YEARS, THE LITTLE DAUGHTER WAS MISSING FROM history.

Mary Johnson Ball Hues, already with a lengthy and long-winded name, did not marry a third time, leaving the little girl with no one she could call father. Sarah Pryor wrote, "The limits of an early colonial house allowed no space for the nursery devoted exclusively to a child and filled with every conceivable appliance for her instruction and amusement. There were no wonderful mechanical animals, life-like in form and color, and capable of exercising many of their functions."[20]

Manners were priority for little girls, partly for future maternal responsibilities. She was responsible for house tasks as deemed fit by her mother, perhaps assisting slaves and servants, and perhaps helping in the gardens of the plantation. It was a typical girl's upbringing.

MARY HUES, FORMERLY MARY BALL, FORMERLY MARY JOHNSON, WROTE her will on December 17, 1720. She died shortly after, in the summer of 1721, as she noted that she was "sick and weak in body."[21]

The person of interest aside, the will itself was remarkably unique. "It is seldom that in a document of this kind," commented historian George Beale, who had accidentally discovered the will in the Northumberland County archives a century ago, that "maternal affection, having other and older children to share its bequests, so concentrates itself upon a youngest daughter, and she a child of thirteen summers."[22] All but two items in the will went to Mary. Among the young Mary's inheritance included, according to the will:

> *One young likely negro woman ... to be delivered unto her ... att [sic] the age of Eighteen years. ... I give and bequeath unto my said Daughter Mary Ball two gold rings the one being a large hoop and the other a stoned Ring ... one young mare and her Increase ... sufficient furniture for the bed her father Joseph Ball left her [including] One suit of good curtains and fallens, one Rugg, one Quilt, one pair Blankets ... one good young Paceing horse together with a good silk plush side-saddle.*

By 1721, Mary Ball, daughter of Joseph and Mary, was fatherless and motherless. She was thirteen years old and she had inherited quite a bit—not enough to be considered rich—at a tender age.[23]

LOSSING CITED A LETTER "IN POSSESSION OF A FRIEND IN BALTIMORE" dated January 14, 1723, from Mary's own handwriting, to her half brother Joseph, the son of her father and his first wife. The fifteen-year-old girl noted,

> *We have not had a school-master in our neighborhood until now in nearly four years. We have now a young minister living with us, who was educated at Oxford, took orders, and came over as assistant to Rev. Kemp, at Gloucester. That parish is too poor to keep both, and he teaches school for his board. He teaches sister Susie and me and Madam Carter's boy and two girls. I am now learning pretty fast. Mama and Susie and I all send love to you and Mary. This from your loving sister, Mary Ball.*[24]

The education of young girls was different from that of the boys and certainly different from that of the men. Governor William Berkeley, the hated leader of Virginia during the 1670s, once wrote, "I thank God, there are no free schools nor printing, and I hope we shall not have these [for a] hundred years; for learning has brought disobedience, and heresy, and sects into the world."[25]

Years after this condemnation of learning backfired, after the establishment of the College of William and Mary and after both public and private tutors became popular, Mary's own half brother, Joseph Jr., wished to educate young men. Court records showed, in 1729, that he submitted a measure of "instructing a certain number of young gentlemen, Virginians born, in the study of divinity, at the county's charge."[26] Of women, though, there was no such measure and no need for one.

"Their destiny was *Kinder, Küche, Kirche*; whether or not they knew German, they knew that," wrote Nancy Turner.[27] Children, kitchen, church. A wholly domestic life.

The schools were few and far between though, as the size and rural setting of Virginia were an impediment. Bishop of London Henry Compton, toward the end of the seventeenth century, noted that "this lack of schools in Virginia is a consequence of their scattered planting."[28] What towns in Virginia in 1700 existed were sporadic.

The most popular way to teach was through a hornbook, a medieval-origin tablet typically with the alphabet and phonemes and the Lord's Prayer. It was also called a horn-gig, a horn-bat, a battledore-book, absey-book, and other names, and its use as a simple primer for children was standard practice in both Europe and colonial America.

Young Mary used these hornbooks to learn to read and write, probably, due to her mother's and stepfather's wealth, with the use of a private tutor, if not partly taught by Mary Hues herself.[29] It was popular a half century later in Pennsylvania and New York to gift them to children, implying that it was effective in providing that rare quality of being both fun and useful.[30] Willard

Randall states that in her studies "she learned what was expected of a Virginia gentlewoman: sewing, dancing, embroidery, the Anglican catechism . . . painting, horseback riding, how to treat her slaves."[31] It was "practical and judicious," meaning numbers, writing, Bible studies.[32] No studies in law or philosophy or science or theology would have been necessary for a young or old woman, being exclusively for the men of those professions.

WHO WAS THIS "SISTER SUSIE" MENTIONED IN THE LETTER? NO SUSAN OR Susie is mentioned in either her father's or mother's will, nor in any primary documentation. It is possible that this was a transcription error from Lossing, reading "Susie" or "sister" for another word, or she used it as a term of endearment for a girlfriend. This letter was reportedly filled with spelling errors and "the handwriting was stiff and cramped," so perhaps it was just misreading.[33] This was not out of the ordinary, however, in a time when standardized English did not exist.

JAMES SHARPLES, AN ENGLISH PAINTER WHO EMIGRATED TO THE NEWLY formed United States in 1794—and who created portraits of such American giants as George Washington, Thomas Jefferson, and Alexander Hamilton—was said to have conversed with the former president at Mount Vernon about his mother.

Here we learn of Thomas Baker, a reverend from England. He was a native of England but sailed to Virginia to the delight of the educated. He kept "up active correspondence with old University chums," author James Walter wrote, second- or thirdhand, "and became a very fountain of European intelligence among the various English families who had turned their backs on the 'old country.'" Sharples noted that George Washington said,

Thomas Baker was a man of refined education, and devoted much of his leisure, of which he had a good deal, in grounding my mother in religious knowledge, which her mind was naturally inclined to receive, so also in directing her studies

in such other branches of instruction as he deemed most fitting and likely to serve her in the education of children. He was in the habit of reading translations of portions of the best classic authors, and which he was very apt in making interesting by contrasts with modern writers. This most excellent man derived very real pleasure in these labours of love, and strove his utmost with, as he was pleased to speak of my mother, "the most amiable, and yet the most impressional character I have ever known, a girl of great personal attractions, and yet utterly unconscious of their possession."

Washington, Sharples claims, continued to speak of an unknown female French tutor also teaching his mother, especially the "rules for female deportment" such as dancing, which suited "my mother's tall and perfect figure."[34]

Of Reverend Baker, we know nothing else. Not his age or skills, though according to one source, he was a mathematician and great-grandfather of English artist Joshua Reynolds.[35] We can infer that Mary Ball learned the tenets of the Anglican Church.

AFTER HER MOTHER'S DEATH, MARY LIVED WITH HER HALF SISTER ELIzabeth and her first husband, Samuel Bonum, whom Elizabeth had married in 1717 or 1718. George Eskridge, the executor of Mary's mother's will, was Samuel's uncle, and lived close to Mary's new, and third, home at a plantation east of Bonum's Creek, Westmoreland County. It was this Samuel, presumably, that Mary honored when naming one of her sons Samuel. The house that they lived in no longer exists, and by the early twentieth century was nothing but a "pile of bricks," but it overlooked a bank of the Potomac River, a setting not rare but still just as extraordinary.

George Eskridge's responsibilities to Mary perhaps extended only to the property, and not to her upbringing, as reported by Paula Felder, as upon Mary Hues's death, the Northumberland court ordered him to "appraise the said decd: Estate in money" and "exhibit an inventory thereof." No mention of tutoring or raising Mary.[36] Nevertheless, many biographies emphasize the personal relationship between the two, often due to the fact that she named her oldest son, to be the future president, after him.

When he was a young law student at the end of the seventeenth century, George Eskridge was said to have been captured by the "Press Gang" in Wales, grouped with forcibly recruited men of the British military, and sold to an unknown planter in Virginia as a servant. "During that time," wrote Sara Pryor, "he was not allowed to communicate with his friends at home. He was treated very harshly, and made to lodge in the kitchen, where he slept." Eight years later, when his "term of service" was over, he left his master, became a lawyer in England, and returned to the Virginia Colony. He had become "eminent among the distinguished citizens of the 'Northern Neck,'" residing in Sandy Point in Westmoreland County.[37]

Tradition both old and new said Mary often visited and lived under George's care at Sandy Point until her marriage ten years later. Sandy Point, only three and a half miles from Elizabeth's house in Bonum's Creek and about twenty-five miles from her future husband's Wakefield plantation, granted Mary many opportunities to ride her inherited horses, sharpening her already considerably talented equestrian skills. She visited the Lee family in Machodoc, to the northwest.

The local church, along with other plantations, manors, and buildings of fertile Westmoreland County and neighboring countrysides, offered a rich route that Mary could ride alone or with George's children, some of whom were close to Mary's own age.

HOW DID THE YOUNG MARY WASHINGTON LOOK? THERE IS NO RELIABLE DEscription of her that exists, yet we can assume that she looked purely English—her long ancestry attested to that. George might have inherited his blue eyes and dark brown hair from his mother. If she was a horseback rider, which is fair to assume considering her inheritance from relatives' wills, her body would have reflected a strong, nimble build of an equestrian.[38]

HER SOCIAL LIFE GROWING UP WAS NOT ONLY CENTERED ON HORSES AND education. One author, Virginia Carmichael, painted a picture:

> *The only social events that took place outside of plantation life were the Governors' entertainments at Williamsburg. But certainly they entertained at home. At Christmas time, there were great festivities at every manor house. Huge fires burned in cavernous fireplaces. Enormous punch bowls graced the side-boards. Holly and mistletoe hung on the walls, and to the music of the fiddle, beaux and belles stept the minuet. In summer there were horse races, cock fights, boat races, bowling games, and tournaments. Wagers were laid on every sport. And the stakes were high.*[39]

Come hell or high water, friends joined friends for events and fun and merriment during the summer, fall, winter, or spring. Dancing was popular for both sexes, and Mary was fond of it, especially as a single maiden among the men. The Potomac and Rappahannock rivers, along with the countless creeks of Virginia, also offered ample opportunities for sailing. These were all activities she would have thought adventurous as she matured.

ON FEBRUARY 22, 1726, SAMUEL BONUM DIED, AND LEFT TO MARY, IN ACCORdance with his will, "my young dapple gray riding horse," a sign of immense respect for the eighteen-year-old sister-in-law.[40] Within five years, she had lost a mother and a brother-in-law, and at about twenty years of age. Her father was long dead.

Within these twenty years of her life, Mary Ball, born in Epping Forest and residing in Westmoreland County, went from house to house, guardian to guardian, growing from a little girl into a woman. James Flexner opined, "With no parents to tame her and as possessor, due to numerous deaths, of a tidy little estate, she became very self-willed."[41]

Yet Nancy Turner said these years "were to prove for Mary Ball the easiest of all her years." She continued, "The times, so far as she was concerned, seemed fairly peaceful. . . . England was taking a holiday from active warfare, though she still kept up her bickering with France over the question of American boundaries. . . ."[42]

The loss of a father, stepfather, mother, and brother-in-law, parents and guardians all, is no small event now, especially at a young age, and it wasn't during the eighteenth century, either. Thus eminent biographer and historian Willard Sterne Randall almost did a disservice to the young maid when he said, matter-of-factly, "Each time a parent or stepparent died, Mary Ball received a legacy in land, livestock, furniture, slaves, cash, and, usually, a good horse."[43] Factually correct; in each inheritance, she received something from a guardian's death. But Randall implied at first viewing that each death was a means to an end: a death occurred; now where's the inheritance, and where's the good horse?

Indeed, the trauma of these losses probably informed Mary's overprotective tendencies toward her eldest son in later life. She had had so many close relatives taken from her. It was understandable that she would wish to keep George out of danger's way, even when such a desire conflicted with his ambition and, ultimately, his role in history.

While the deaths of Mary's loved ones and family should not be trivialized, Turner was not incorrect in saying these were quiet years. The colonies in the early 1700s were stable, finally forming after a rocky and often bloody and disastrous start. England had two monarchs in these years, switching a dynasty from the Stuarts to the Hanovers with relative ease, at least compared to the previous two centuries. England was rich, the emperor of the sea with its mighty fleets, as evidenced by the colonies.

Mary had an entire life, and family, and wars, ahead of her.

ALEXIS DE TOCQUEVILLE, THE GREAT NINETEENTH-CENTURY FRENCH diplomat, visited the United States in 1831 to study American culture. After returning, he wrote *Democracy in America*, his two-volume masterpiece explaining the differences between

nineteenth-century French and American culture—one aristocratic, another democratic; one with a history of nobility and
kings and queens, another with presidents and representatives.

He wrote of the typical nineteenth-century American girl:

*She has scarcely ceased to be a child when she already thinks
for herself, speaks with freedom, and acts on her own impulse.
The great scene of the world is constantly open to her view; far
from seeking concealment, it is every day disclosed to her more
completely, and she is taught to survey it with a firm and calm
gaze. Thus the vices and dangers of society are early revealed to
her; as she sees them clearly, she views them without illusions,
and braves them without fear; for she is full of reliance on her
own strength, and her reliance seems to be shared by all who are
about her. An American girl scarcely ever displays that virginal
bloom in the midst of young desires, or that innocent and ingen-
uous grace which usually attends the European woman in the
transition from girlhood to youth. It is rarely that an American
woman at any age displays childish timidity or ignorance. Like
the young women of Europe, she seeks to please, but she knows
precisely the cost of pleasing. If she does not abandon herself
to evil, at least she knows that it exists; and she is remarkable
rather for purity of manners than for chastity of mind. I have
been frequently surprised, and almost frightened, at the singular address and happy boldness with which young women in
America contrive to manage their thoughts and their language
amidst all the difficulties of stimulating conversation; a philosopher would have stumbled at every step along the narrow path
which they trod without accidents and without effort. It is easy
indeed to perceive that, even amidst the independence of early
youth, an American woman is always mistress of herself; she
indulges in all permitted pleasures, without yielding herself up
to any of them; and her reason never allows the reins of self-
guidance to drop, though it often seems to hold them loosely.[44]*

It could apply to the girl in the 1830s, and it could apply to the girl in the 1700s.

A century had passed between Mary Ball's girlhood and Tocqueville's writing, in which the idea of the United States, separate from the English Crown, was not even toyed with.

The two were nearly indistinguishable.

Chapter 4

The Marriage of Mary Ball and Augustine Washington

1731

"Wilt thou have this man to thy wedded husband...."

No one alive today knows for sure what Mary Ball Washington looked like. She was once described by a great-granddaughter as being "of medium size, and well-proportioned, the dignity of bearing and the erect carriage giving something of stateliness to her presence, while her brown hair was fine, and her eyes a clear blue."[1] Eminent historian Benson John Lossing elaborated, "She was of the full average height of women, and in person she was compactly built and well proportioned. She possessed great physical strength and powers of endurance, and enjoyed through life robust health." Writing in the late 1800s, he presumably heard of this straight from direct relatives or others in the Lancaster and Fredericksburg, Virginia, region who knew her or knew of her. "Her features were strongly marked, but pleasing in expression," he continued. "At the same time, there was a dignity in her manner that was at first somewhat repellant to a stranger, but it always commanded the most thorough respect from her friends and acquaintances. Her voice was sweet, almost musical in its cadences, yet it was firm and decided, and she was always cheerful in spirit."[2] No doubt the memories of a family member would be glossy and protective. Other descriptions of Mary vary greatly from this one.

These descriptions of Mary were all passed down through folklore, and no known original portrait was painted of her while she lived. The only contender was a somewhat mysterious 18-by-21-inch painting of an elderly woman "at the age of about four-score" in profile, painted supposedly around 1786 by Robert Edge Pine. A known painter, Pine had created such works as *Congress Voting Independence* around 1784 and a likeness of George Washington in 1785.

This portrait was discovered in Fredericksburg in 1850, and through the following decades was passed on until 1916, when a photomechanical print of the portrait was made. At the bottom of the print, as inscribed by W. Lanier Washington, a descendant of Mary's, was written, "There is no other portrait of Washington's mother so well authenticated as this. However, it is still almost unknown to the American people, and nearly all of those who see it in this copy are seeing it for the first time."[3]

It is unclear how truthful the printer's claim was, and experts today cast doubt on its authenticity. Art expert and historian Charles Henry Hart in the early twentieth century stated that there was "not any direct evidence" of its legitimacy.[4] Recent scholarship has also doubted its authenticity. The website of Mount Vernon has said that the portrait is "spurious"[5] and doubts have been cast on its true depiction, possibly due to its questionable origins and absence from public record. Executive director Karen Hart of the Mary Ball Washington Museum and Library in Lancaster, Virginia, stated that "there is no evidence that Mary Ball Washington ever sat for a portrait. . . . These works of art are based on a mix of truth and myth."[6] Michelle Hamilton, the manager of the Mary Washington House in Fredericksburg, Virginia, also waved it away, calling it nothing more than a "feminized version of George" Washington.

"That's not her," she definitively declared.[7]

Yet another, George Washington Parke Custis, Mary's own great-grandson, confirmed "there was no portrait extant of the Mother of Washington." Her own grandson would have known of the existence of even a single painting of his grandmother, which at the least would have been wanted—if not "highly prized"—by Washington himself.[8] George Washington Parke Custis's sister, Nelly Lewis, married to Betty and Fielding's son Lawrence, agreed similarly, arguing, "I do not believe the [general's] mother ever had her likeness taken by any one—and certainly if it ever had been taken, her *children* and strangers would have possessed it."[9]

What we do know is that something about her caught the eye of Augustine Washington.

WHILE MARY GREW FROM CHILDHOOD TO WOMANHOOD IN EARLY eighteenth-century Westmoreland County, Augustine Washington, born 1694, over a decade her senior, was also learning and maturing. Like Mary, he experienced tragic losses in his youth, a shared experience which might have brought them together. On the other hand, Augustine had lived a lot of life by the time he encountered young Mary Ball.

The future husband of Mary Ball lost his father, Lawrence, when he was only four years old, mirroring Mary's loss of her own father at a young age. Under the care of his mother, he learned typical Virginian boyhood tasks befitting "the well-born, well-endowed colonial youths of the period." This included military training, the hunting of all sorts of animals and wildlife, and management of plantations and properties.[10]

Mildred Washington remarried around 1700 to rich merchant George Gale of Whitehaven, Cumberland, England, to which they, Augustine and his siblings included, relocated soon after. They had met in Virginia, as George's profession required constant trading between the colonies and England. Soon after the move, George and Mildred had a child, a daughter also named Mildred, on January 25, 1701, but she died two months later.

The mother Mildred, however, was spared the pain of seeing her own daughter die in infancy, as she died shortly thereafter and was buried five days after the birth of her daughter. Mildred's will, made shortly before her death, noted that she was "doubtful of the recovery of my present sickness," and left part of her inheritance from her previous husband to his brother, John Washington, and another to her current husband. To George Gale she also gave the right to raise her children.

She was buried at the Chapel of Saint Nicholas in Whitehaven, a small, uncomplicated church. Today, a plaque marks the cemetery. (The Union Jack and thirteen-starred Stars and Stripes are also displayed, crossed in union[11]).

Augustine had lost his father and mother by the time he was seven years old. Under his stepfather's supervision, he was educated at Appleby School in the very north of England, in the county of Westmorland. Sometime later—not much later, within three years of Mildred's death—John Washington, a cousin to Augustine's father, won custody of the children, the ten-year-old included.[12]

They went back to Virginia in 1704.

RON CHERNOW WROTE, "RAW-BONED AND GOOD-NATURED, AUGUSTINE WASHington remains a shadowy figure in the family saga."[13]

The next decade in young Augustine's life, from 1704 to 1713, was marked by the guardianship of his father's cousin, who raised him along with his own children, all living in Chotank, Stafford County, Virginia. It was close to Bridges Creek. A receipt for all of George Gale's responsibilities and guardianship was signed on April 6, 1704, to "hereby discharge the said George Gayle from all further demands on account of the Estates and portions of the said children."[14]

Augustine was called "Gus" by those who knew him, particularly friends, and had an ambitious personality suiting any Virginian wanting land.[15] He acquired land in 1715 when he came of age, at twenty-one. Included in this inheritance were the nearby

Bridges Creek plantation; over a thousand acres of fertile land; to-bacco, slaves, horses, sheep, and livestock, and many other house-hold or plantation items such as dishes, two dogs, tools, saddles, and kettles.[16]

That same year, 1715, he married Jane Butler, five years younger, the daughter of neighbor Caleb Butler. This was the first of his "several Ventures," he wrote years later in his will.[17] Perhaps that was with a touch of humor or with some lighthearted fun about the relationship between himself and his wives; it also offered more of who he was than anything like deeds and inheritance records. But with Jane Butler came 1,750 acres of land, which was combined with his inheritance at Bridges Creek.[18]

Through these fourteen years of marriage with Jane Washing-ton (née Butler) from 1715 to 1729 the couple had four children: Butler, Lawrence, Augustine Jr., and Jane. Butler died in infancy, and Jane died in childhood.[19] Lawrence Washington, in his later years, would prove to be the best mentor to his younger half brother George, influencing him more than perhaps any other man in the boy's life.

THESE YEARS FOR AUGUSTINE WERE RICH IN MONEY, WEALTH, AND INFLU-ence, as he became more aware of his interest and talents, more aware of himself. In these years he acquired land; contracts and grants and leases; and court orders to right what he perceived as wrongs. In the court orders of Westmoreland County, Au-gustine Washington filed suit against a certain William Brown Mulatto, and "declared against him for the sum of 794 pds of tobacco." Brown did not appear before the Westmoreland court twice, and thus on May 28, 1725, his absence "confirmed" the debt.[20]

OVER 1,700 ACRES BELONGED TO THE NEWLY MARRIED AUGUSTINE, AND HE purchased more in 1718 at Popes Creek from Joseph Abbington, already expanding his ancestors' property, for a large sum of 280 pounds. It was a cleared land, having been used for decades before.[21] Four years later, in 1722, he continued construction on the house, completed by 1728.

It was at this home at Popes Creek that Augustine would spend about a decade of his life, nearly the longest he continuously stayed in a single home. Additional plantations and farms were bought and conjoined. It was at this home that he and his second wife had children (all his children from his first were born before the house was completed. This home is now marked with a federally owned national park monument.)

THE REGION WAS RICH IN IRON, A HIGHLY DESIRED COMMODITY FOR THE Mother Country. With that in mind, in 1723, John England at the Principio Company visited the region, and soon Augustine made a deal with the English mining company to lease out part of his land. An agreement between the two was signed by March 2, 1729, stating that three years earlier, on July 24, 1726, Augustine agreed to "demise, grant, bargain, and sell" about 1,600 acres of land to mine iron ore.[22] He would be compensated and share a portion of the profit for the use of his land. "This," says Charles Hatch, "turned Augustine in a new direction as he became particularly interested in iron manufacturing in its various aspects—financial, managerial and operational."[23] An iron furnace was built on his land in Stafford County, and Principio soon after owned land in Maryland, becoming one of the most significant iron producers in colonial America.

The move would prove profitable both to Augustine, who would eventually come to own about a twelfth of the company, and to Principio itself, as Virginia pig-iron exports by the mid-1700s were over three thousand tons per annum, ten times more than the rest of the colonies. To compare, all of England's native iron amounted to less than seventeen thousand tons per year.[24] The contract stated, in addition to the leased land, that Augustine would receive 20 shillings per ton carted from the mine to the furnace, about two miles, which required additional workers, servants, and slaves.[25]

To be clear, it was not an easy task leasing the land or contracting for iron. Douglas Freeman noted as much, as "in general, the Virginia iron industry was not prosperous. Operators of colonial furnaces were not permitted under British law to export bar iron. Nor were they allowed to make iron castings, such as pots, firebacks and andirons," though there are instances of owners selling such items in reality. Freeman continues that in Augustine's case specifically, the contract was "an acceptable one," but it was straining.[26]

Augustine's land owned by Principio Company was later designated a historical archaeological site, entering the National Register of Historic Places in 1984.

JANE WASHINGTON DIED NOVEMBER 24, 1729. HER HUSBAND WAS NOT BY HER side, but in England negotiating an additional contract with Principio. In fact, upon learning of his wife's death when he returned, he froze . . . in productivity, in interest in money, in land, in leasing. He became demotivated, deflated. It was so apparent he was in grief that one stockholder of Principio Company, John Wightwick, wrote on October 2, 1730, to John England, "I think this is treating us neither kindly nor honorably to leave us in so great a state of uncertainty." Augustine refused to answer questions or make decisions that were laid out to him, despite "having sufficient time to consider it," according to the letter.[27]

He was alone, with three children and no wife to mother them.

Congruent to all this, during these same years, young Mary Ball in the Northern Neck had lost her father, mother, and stepfather, and had grown up under the wing of her half sister Elizabeth and her husband.

THE ADULT AUGUSTINE WAS SAID TO BE OF HANDSOME STATURE, SIX FEET tall and muscular, with "the most manly proportions," as reported by George Washington Parke Custis. Augustine was a gentle giant who could "raise up and place in a wagon a mass of iron that two ordinary men could barely lift from the ground." This feat alone allowed him to become a sort of Hercules of Virginia; no man wished to duel or combat him lest they lose.[28] It all sounded very mythical, a man with godlike super strength.

Ella Bassett Washington said that Augustine was "a noble-looking man, of distinguished bearing, tall and athletic, with fair, florid complexion, brown hair, and fine gray eyes."[29] He had a handsome physical appearance, which perhaps made Mary more attracted to him than to other bachelors. Families of their stock in Virginia knew one another, especially those who had been in the New World for years before. Her first meeting with Augustine could have been when he was still married to Jane. They may have also met at the home of George Eskridge, Mary's guardian, tying everything to a single place.[30]

(That is why it is curious to see some legends that the two met in England, while she was visiting her half brother Joseph, Mary marrying Augustine in Cookham, a small village along the River Thames in southeast England. The story goes, almost as if written by a great romantic author, that Augustine injured himself when his carriage crashed, and Mary, whose gate at which he crashed, took him and nursed him back to health.[31] But even Mary Virginia Terhune (writing as Marion Harland), who was sometimes keen to emphasize the poetic and the romantic tales over known facts, was skeptical of it.[32] There is no proof for it; in fact, there is proof against it. Joseph was in Virginia at that time, so no relative of Mary's would have stayed in the Old World. The time frame was much too tight for Augustine to visit England after Jane's death and meet Mary, marry her, have George, and return to Virginia—and make no mark on anyone's lives.)

Mary Ball was older for an unmarried woman, in her early twenties at this point. Could she have been content with an unmarried lifestyle? Could it have been because of a fiery (or, to some, deficient) personality that turned men off? Certainly not every woman would have to be married, but as the only daughter—much less only child—of her mother and father, it would have seemed an odd choice for a Ball. She missed "by the narrowest margin being classified as an old maid." But luck determined the Washington-Ball bloodline would live on.[33]

Some biographies take Mary Ball's relatively old age in marriage as proof of her strength of character: "Could it have been," opined writer Willard Randall, "that there was something so strong and independent about her that every suitor seemed to back away?"[34] Given Mary's famous stubbornness in later life, and her singular courage in running a plantation on her own, it's possible that many suitors found her too fearsome for their taste, at least at first.

She would have been defying considerable social pressure to marry; being of old age and unmarried was almost insulting. The thought of an old unmarried woman was enough to almost cause slander. A 1772 article from the *Virginia Gazette* talked about an unmarried woman of fifty-five:

> *Mrs. Mary Morgan has lived to the Age of fifty five unmarried, but she merits no Blame on Account of her Virginity, for she certainly would have entered into the Marriage State if any Man had thought proper to make his Addresses to her. Nature has bestowed on her no Beauty, and not much Sweetness of Temper; the Sight of every pretty Woman, therefore, is very offensive to her, the Sight of a married One hardly supportable.... Inwardly tortured by her own ill Nature, she is incapable of any Satisfaction but what arises from teasing others....*[35]

Though the difference in age was generational between Mary Ball and Mary Morgan, the sentiment is clear: an unmarried woman, especially an *old* unmarried woman, was unsatisfying, mean, decrepit, and unwanted. Mary Ball, if she remained unmarried, would have been swept aside by history.

NO ONE KNOWS WHEN AUGUSTINE FIRST MET MARY, OR WHEN HE ASKED FOR her hand in marriage, or when he first fell in love with her.

Mary and Augustine were married on March 6, 1731.[36] Mary was about twenty-three years old; Augustine, thirty-seven.

Where they were married was omitted in the family Bible. Theories exist, placing the marriage all around the area, from Mary's girlhood church, Yeocomico,[37] to her home at Sandy Point, with Cople Parish's Reverend Walter Jones officiating.[38] Both were valid options and there is an equal possibility that both occurred, though a typical wedding in Virginia in the mid-1700s was at home, in the late morning. Wherever it did take place, it was an event for the ages. Three times, in the three preceding weeks, the "banns" of marriage—an announcement for the upcoming celebration and ceremony—were published in pamphlets and papers.

MARY BALL WASHINGTON WAS NOW THE WIFE OF HER HUSBAND AUGUSTINE, and Augustine Washington was the husband of his wife.

Marriage was not simply a means to an end, to secure a fortune or dowry or inheritance, though that was certainly the case for many. It was also seen as a legitimate bond of love. There is no reason that this would be different between Mary and Augustine, though the latter had ample practical reason to bring another to the house, in order to manage the plantation. Mary and Augustine may well have been a solid married couple, happy with each other. "Spouses frequently expressed feelings of close, tender regard for each other, suggesting that the romantic grandiloquence characteristic of courtship contained more than a small portion of genuine love."[39] When a husband or wife went away for family or business, it left a hole in the heart and household of the plantation.[40] In the decade following, Augustine frequently visited England for business, leaving Mary alone with her children.

This did not mean that Mary and Augustine were equal members of the household. Augustine was the head. There was a clear role for the sexes—and for sex itself. "The fragmentary evidence available on marital sexual behavior," said Daniel Smith, "suggests that while husbands and wives enjoyed and needed sexual advances, only men were expected to make the advances."[41] Mary ended up having six children with Augustine, and unknown to history is how many miscarriages she may have had.

THEY DID HAVE CONJUGAL RELATIONS, AND QUICKLY. IT DID NOT TAKE LONG for Mary to become pregnant. A mere eleven months after her wedding, she gave birth to her first child. These eleven months are spent in historical obscurity, filled with legends more fit for bedtime stories than history. She lived at Popes Creek, with her belongings coming with her in the marriage. This was her fifth home, after her birth home Epping Forest, her home with her stepfather George Gale at Cherry Point, her home with George Eskridge at Sandy Point, and her home with her half sister Elizabeth at Bonum's Creek.

It was during these eleven months that Mary was hosting a friend at her house. There was a dark, thunderous storm outside that evening, and supper was well under way when a bolt of lightning struck the house. It traveled downward, struck Mary's guest, and killed her instantly. The bolt was so direct that it fused the guest's flesh and utensils together. Mary Terhune recalled the story: "The nervous shock left ineffaceable traces upon the strong mind. . . . She grew pale and sick at the approach of a thunderstorm, and at the first roll and gleam of the deadly elements sought her own room or sat with closed eyes and folded hands, absorbed in silent prayer while it lasted."[42] George Washington Parke Custis, from whence this tale came, said this was her singular fear: lightning.[43] But with this fear came more problems.

Traumatic events of any sort could haunt people the rest of their lives; fatal storms that killed dinner companions more so. The cause of lightning was unknown in the 1730s, with some scientists dabbling in theories of its origins. (Benjamin Franklin's famed kite experiment did not occur until 1752, much later.) To the scared Mary, who would not have been interested in or been exposed to the theoretical reasons of electricity and lightning, a sudden strike of literal death from the sky would have shaken her to the core. Whether from the wrath of God for some misdeed or simple bad luck that God allowed, something happened that in all likelihood would have left a mark.

This event could have been the singular moment when Mary

would have gone from being a normal prospective mother to an overbearing and overprotective prospective mother. There is no *primary* evidence that it happened (Douglas Freeman calls the incident spurious, though admits "some actual unpleasant occurrences may underlie the sensational yarn"[44]), but it certainly fits the growth of her character. The thunderstorm, or whatever may or may not have happened that night, shaped the pregnant Mary from her childhood innocence to something more, something that would similarly shape the life of her son. Something during her pregnancy—it could have been the thunder, it could have been another brush with death—changed her.

This death of a friend may have very well shaped how she viewed the world: a world where her mother and father and several siblings were dead, a world where death will come with but a moment's notice. What she may have seen as protective, others, even today, may see as helicopter-parenting and mean. It's a tale as old as time, the struggle to find the perfect balance of parental worry and youthful independence.

One account tells us this fear of thunder and lightning remained with her for life. In one instance decades later, her daughter Betty, during a particularly bad storm, caught Mary with her head down at the bed, hands clasped in prayer. When Mary saw her daughter, she reportedly said, "I have been striving for years against this weakness, for you know, Betty, my trust is in God; but sometimes my fears are stronger than my faith."[45]

She would indeed have many, many fears in the years to come. For herself and her husband and her children and her very country.

ON A COLD WINTER DAY IN VIRGINIA, AT THE PLANTATION ON POPES CREEK, was born their first son. He was named George, in honor of the guardian who so helped his mother during her youth. No physical description of him exists, but he must have been a healthy, plump baby to survive at such a period with such strength into his adulthood. He was born, according to the Washington Bible, on February 11, 1731 or 1732 (it reads "1731/2"). The former year is correct.

It noted that he was born "about 10 in the morning." Nearly a century later, his step-grandson, George Washington Parke Custis, wished to memorialize this spot. In 1815, traveling in his private schooner *Lady of the Lake* to Westmoreland, he landed near Popes Creek, with a "slab of freestone," bearing the engraved:

HERE

THE 11TH OF FEBRUARY, 1732,

GEORGE WASHINGTON

WAS BORN

With him was a relative of Washington, Samuel Lewis, along with the schooner's crew and others, who were "desirous of making the [ceremony] of depositing the stone as imposing as circumstances would permit." He later wrote in a letter to the *Arlington Gazette* editor, "enveloped it in the 'star-spangled banner' of our country." They placed the stone near an ancient and decrepit chimney of George Washington's former childhood home and departed with a cannon salute.[46]

The slab was worthy of any relic of a saint, wrapped in the righteous cloth of the American flag, and a ceremony was held as reverently as any liturgy and benediction praising God. But this was not for the birth of the savior of mankind millennia ago in Bethlehem, Israel, near the mighty Jordan River; this was for the birth of the savior of a nation in Popes Creek, Virginia, near the mighty Potomac River, a mere eighty-three years earlier.

Unfortunately, this American relic was "in fragments" by about 1857 and was reported missing about 1870, and has not been seen since.[47] Furthermore, and unfortunately, it contains a glaring error: the year of Washington's birth.

The discrepancy can easily be explained. Most of Western Europe and its colonies were under the Gregorian calendar, created by Pope Gregory XIII in 1582 to accurately reflect the original date of the Resurrection of Christ and its corresponding spring equinox. The calendar used before, called the Julian calendar after Julius Caesar, contained an error in which, centuries later, the date of the equinox changed considerably. The papal bull that Gregory issued did not extend beyond Catholic countries, and thus did not reach England, lest the Crown, in their view, accept the supremacy of the pope. Today, all but four countries in the world use the Gregorian calendar, while most Eastern Orthodox churches continue to use the Julian calendar in their liturgy.

By 1752, over 170 years after its initial inception, England and its territories and colonies had to change from the Old Style. Two years earlier, the British Parliament passed the Calendar (New Style) Act, designating January 1 the new year, a change from March 25, and adding an additional eleven days to the calendar. Wednesday, September 2, 1752, was followed by Thursday, September 14, 1752. Additionally, in order to account for the change of New Year's Day, all dates between January 1 and March 25 had to add one year. George Washington Parke Custis accounted for the latter by adding a year to the plaque, not the former by adding an additional eleven days. Thus he engraved a sort of combination of Old Style–New Style date of birth, being wrong in both.

Time and date and year changes were confusing enough, but calculating these did not even matter to George Washington. To him and to his contemporaries, his birthday was and continued to be February 11 . . . most times. The Comte de Rochambeau Jean-Baptiste Donatien de Vimeur, a French ally during the Revolutionary War, wrote to Commander Washington on February 12, 1781, saying, "We have put off celebrating that holiday till to-day [instead of yesterday], by reason of the Lord's day and we will celebrate it with the sole regret that your Excellency be not a Witness of the effusion and gladness of our hearts."[48] Henry Knox noted his birthday on February 11, 1790, and on February 12, 1798, George "went with the family to a Ball in Alexa[ndria, Virginia]" in celebration.[49] On the other hand, a poem written by Elizabeth Willing Powel in 1792 commemorates the president's birth on February 22.

But that was all decades ahead. The date of February 11, 1731 was George Washington's unchallenged date of birth for the time being.

AT THE TIME, A BIRTH WAS JUST ANOTHER DAY IN THE LIFE, ESPECIALLY FOR Augustine, George's father. "People altered so little in those slowly shifting times that in all probability at noon of that amazing morning Augustine Washington was every whit as calm" as could be, maybe could "even have saddled his horse and ridden off, as the others did."[50] Indeed, this is not an exaggeration—it was a way of life, as seen in Joshua Hempstead of Connecticut's diary, in which he casually writes on July 30, 1716, "I was all day Getting Waylogs & Launching Timber [and] my wife delivered of a Daughter about Sunset." The next day, he went back to work, making a maiden voyage of his sloop ship.[51] Another man, Richard Lloyd from Maryland, did not know of his daughter's birth until a house servant told him.[52]

Yes, George was his first son of this marriage, but Augustine already had several other children from his first marriage, some blooming into young adulthood, who were just as legitimate and just as "Washington" as George himself. He had priorities to raise them and raise the crops and livestock and business of his plantation. That is not to say there was an indifference to the birth of George. Life was simply, for the father, a day-to-day tackle of business and rearing of the maturing children.

If the birth of a child was mundane to Joshua Hempstead or even Augustine Washington, the mother's feelings were perhaps the exact opposite. Childbirth is scary and painful enough with modern medicine, and in the eighteenth century, at home with only a comforting midwife and maybe some female friends or relatives, the fear escalated. It was rare that a doctor was present, and even rarer if the husband was. One late eighteenth-century birth in Maryland was described as "confusion & distraction," and one woman saw her sister-in-law's delivery as a "Scene of Sickness" where there was "so much complaining, and [she was] so low spirited in her lying-inn."[53]

"THE ATHENS OF VIRGINIA"—THAT WAS THE TITLE BISHOP WILLIAM MEADE gave to the place of George Washington's birth.[54]

The house burned to the ground on Christmas Day, 1779. It was probably two stories tall, with four rooms on each, with a steep roof and large chimney of brick or stone, the same that Custis visited.[55] Most if not all of it was made of brick, according to excavations. According to Benson Lossing in 1871, "The only approach to ornament was a Dutch-tiled chimney-piece in the best room, covered with rude pictures of Scriptural scenes." Included in his work is a drawing of a circular Dutch tile, perhaps as an example and not as a specific one owned by the family, of a leviathan-like creature swallowing Jonah.[56] Another source, writing in 1895, states that the house was "very comfortable" and "not a great nor a grand house," with a chimney on each end, no carpets or oil, and few books.[57] Sara Pryor states this is partly an embellishment of historians to imagine and "think that their great men were cradled in poverty." How unpresuming they must have been to be raised in a small house, to become indispensable to a nation from such humble origins! While the house at Popes Creek was not extravagant, it was not small, but a "larger . . . modest dwelling." Pryor continued, "The universal plan of the Virginia house of 1740 included four rooms, divided by a central 'passage' (never called a 'hall') running from front to rear and used as the summer sitting room of the family. From this a short staircase ascended to dormer-windowed rooms above." As more children came and the family grew, so too did the house, with extra rooms built near the chimney "without regard to architectural effort."[58] The inventory of the house in the 1760s included ten bedframes, thirteen tables, nearly sixty chairs, and eight mirrors and chests—not exactly a pauper's house, it was.[59]

The view itself was spectacular, and purely American, however, with an overlook to the grand Potomac River, the shore of colonial Maryland visible on the other side. It was surrounded by fig trees and vines, shrubs, and flowers.[60] The house was mounted on a hill twenty feet in elevation, allowing surrounding fields and marshes to dominate. A farm road going northwest passed the family cemetery, where George's grandfather and great-grandfather were buried, eventually ending at the Potomac River itself.

HIS BIRTHPLACE HAS BECOME ALMOST A SHRINE FOR THE AMERICAN RELIgion. Indeed, James Thomas Flexner noted that had Washington's birth "taken place in classical times, surely some prodigy would have been initiated from Olympus: a snake sent down to be strangled by the hero in his cradle, or at least a weird clap of thunder."[61] Centuries later, many see George Washington's birth as a spiritual day of observance. In commemoration of the centennial of his birth, sculptor Horatio Greenough, commissioned by the United States government, created a massive eleven-foot-high sculpture of Washington, right arm raised and finger extended upward. He sits on a chair—or a throne, depending on one's point of view of the first president. In his left hand is a sheathed sword. It's classical Roman and classical Greek in style, looking more like Zeus or a sculpture of a Roman emperor. The imposing larger-than-life sculpture sits overlooking the west end of the west wing of the National Museum of American History.

MARY WASHINGTON HAD FIVE MORE CHILDREN WITH AUGUSTINE, SOMEtimes with hardly a break. George was born in early 1732, according to the New Style of the calendar. In 1733 came his sister, Elizabeth "Betty" Washington, who married Fielding Lewis in 1750. In late 1734 came Samuel Washington, then John Augustine in 1736. Charles Washington, George's youngest brother, was born in May of 1738. Elizabeth and Samuel were also names of the Eskridge clan and relatives of George; it can be presumed, if she named her firstborn after George himself, that she named her others after the wife and children of Eskridge as well, if not her own siblings.

Finally, there was Mildred, born June 21, 1739. She died a mere infant, at sixteen months of age on October 23, 1740.

With each pregnancy came the risk of childhood mortality. George, or any of the children, could have died before even seeing the light of day, whether through complications of birth or being stillborn or miscarried. In 1771, one of the twins born to a couple in the Chesapeake region "hanged himself in the navelstring." Landon Carter, the son of Robert "King" Carter of Virginia, saw everyone in his house "in a great fright and [the mother, Landon's daughter-in-law] almost in dispair [sic]. The child was dead and the womb was fallen down and what not."[62] This easily could have occurred in the Washington household.

Likewise, Mary herself could have simply passed away, leaving Augustine a widower once again. The high death rate among childbearing women could have easily killed Mary in these seven years, with all the pregnancies being so close together.

MARY WAS NO LONGER A BALL—SHE WAS NOW A WASHINGTON. SHE WAS NO longer a maiden—she was a mother. With her marriage and the birth of her first child, Mary's life entered a new phase and new chapter that, though common among all women, nonetheless affected her. And though her marriage to Augustine lasted twelve years—he died in 1743—this was a life-changing and history-changing union. We know of Mary not as a maiden Ball or the "Rose of Epping Forest"—but as a mother, of George Washington.

And that changed everything.

Chapter 5

In the Shadow of the Empire

THE EARLY MOTHERHOOD OF MARY BALL WASHINGTON

1732–1738

"A baby of unusual heft."

What would it have been like if George Washington was born not in Virginia, but in his ancestral England? For a time, some thought he was. Pamphlets had been written, claiming that the Father of America was in fact English, placing a greater emphasis on breaking with his native land.[1] This was a direct contradiction of a vast majority of accounts, and what proof they offered was shortly swept away. George Washington Parke Custis wrote a letter in 1851 to the *Boston Evening Transcript* in response to those who claimed he was born on English soil. He held nothing back: "Lord Byron wrote of an age of bronze, but we live in an age of brass; for surely the very idea that Washington was born in England is too monstrous an absurdity to be brazened to the world in the nineteenth century." He considered it insulting to both his step-grandfather and his country to even suggest such a thing.[2] Parson Weems, who brought to us the myths of chopping down cherry trees in the first hagiographical biography of the first president, shrugged off the idea as well, arguing that George Washington was great not because of his false English birth, but because of his actual American birth.[3]

Still, it raises the question: How would George have grown? For one, his education may have been phenomenally different.

The education in, say, London would been much more theoretical than what George actually received in Virginia, perhaps dwelling on law or theology. Plantation management and surveying were more of an emphasis in the colonies and not in Old England.

More important, an English-born George would have, perhaps, been raised in an English culture, not a colonial culture, where the troubles of taxes and Indians and French colonies were only a faraway problem, not a crisis experienced firsthand.

All that said, George did spend his first few years very much under the influence of the British Empire. He lived in Britain's prize colony, attended the Anglican church, and was raised in a world defined by his privilege and pedigree.

AS IT IS, GEORGE NEVER DID WRITE ABOUT HIS EARLY LIFE. IF HE HAD, WE might have been reading even now of his infancy and childhood with colorful, sharp, and personal anecdotes that would have brought realism to the otherwise larger-than-life man. Mary's influence during these years was probably more spiritual than practical, as she guided her children's religious upbringing. The first years of George's life would invariably affect his outlook on the world, but most important his character. In John Stevens Cabot Abbott's biography of Washington, he wrote, "Trained by such parents, and in such a home, George, from infancy, developed a noble character. . . . Happiness in childhood is one of the most essential elements in the formation of a good character. This child had before him the example of all domestic and Christian virtues."[4]

The closest thing we have to an account of Washington's childhood is the long-ago authorized biography from David Humphreys, an aide-de-camp and friend of Washington. Still, this work had some factual errors and gaps, such as Washington's own birthday and year, which was cited as 1734. Several hints were given to George's childhood in it, though: He was raised with a typical eighteenth-century education for the upper class. He learned numbers, geometry, mathematics, geography, history, and humanities. Very early on, he became interested in dancing, fencing, and horseback riding—and military exercises were of particular interest. It was typical for a boy of his age, extending into the teenage years.[5] It was not remarkably different from any other planter's childhood. The first twenty years of his life in Humphreys's biography got no more than five paragraphs. For a man about whom many want to know every facet of life, they are left teased with generalities.

Compared to, say, the youth of Benjamin Franklin, a wordsmith and reader himself, who devoted several chapters of his autobiography to his own childhood—describing his relationship to his mother and father and siblings, as well as his love of reading and education—Washington's childhood is lacking in detail in written works. The Father of His Country never wrote an autobiography.

Born over twenty years before George, in 1706 in Boston, Franklin in his own works offered an insight into his childhood. "From a child I was fond of reading, and all the little money that came into my hands was ever laid out in books," he had written. "Pleased with the *Pilgrim's Progress*, my first collection was of John Bunyan's works in separate little volumes."[6]

However, differences in age and location were fundamental enough that one can't compare the childhood of one Founding Father, such as Franklin, to that of George Washington.

Perhaps other autobiographies, such as that of John Adams, born October 1735, only three years after George, would work. But again, the cultural differences of Massachusetts, with its Puritan emphasis and background, could not be compared to rural, Anglican Virginia. Adams's profession as a lawyer—and he was a graduate of Harvard University, to boot—also was radically different from Washington's background.

Perhaps the life of Thomas Jefferson, born a relatively few eleven years after George, near Charlottesville, Virginia, would be more appropriate in terms of piecing together the life of an early Virginian infant to a Virginian plantation owner. He—Jefferson—was born in April of 1743 to Peter and Jane Jefferson, who had ten children, and by the age of five was taught by Reverend William Douglas of Saint James Northam Parish, Goochland County, Virginia. By age nine, he was taught Latin, the lingua franca of the educated world. Jefferson's father died when he was fourteen.[7]

In many ways Thomas and George, contemporaries, grew up in similar households. Their fathers died when they were teenagers, and they were left with plantations in rural Virginia. Yet still we see a diverging commonality during their childhoods, and no two Founding Fathers—no matter how close in age or place—could accurately be compared.

Neither Augustine Washington nor Mary Washington kept diaries. Letters of contemporaries certainly existed, sharing a human element for many of the families, but, again, it was difficult to re-create these early years of George's life.

BEFORE HIS FATHER'S DEATH, BEFORE THEIR MOVE TO LITTLE HUNTING Creek, before their move to Fredericksburg, before much of what defined George Washington as George Washington—his adolescence, his striving for independence—there was the newborn George, firstborn child of Mary and Augustine Washington.

He was described as "a baby of unusual heft," according to Ron Chernow.[8] These first years were the most vulnerable for the young George, as was true of all babies of this era. Diseases all but eradicated today were rampant and dangerous to the immune system of the young. In the eighteenth century, one-tenth of infants died before they were a year old, and two-fifths died before they were six.[9] (To compare, infant mortality in the United States today is less than six deaths per thousand births, or less than 0.6 percent.[10]) But the average was sometimes more disastrous in reality, and statistics changed from place to place and from location to location. Samuel Sewall, known primarily as one of the judges during the Salem witch trials in the late 1600s, and his first wife, Hannah, saw seven of their fourteen children die before they were two, and another was stillborn. Another New

England woman, Mary Vial Holyoke, who married in 1759, had twelve children in the first twenty-three years of marriage—she was pregnant for about a third of these years. Eight of her children died before they were three, with all but one dying at birth or infancy. "Mary Holyoke had more pregnancies and suffered more child deaths than her average contemporary, but her story presents a poignant example of the extreme physical trials some women endured."[11] Indeed, Mary Washington's last child was one of those claimed in death during infancy as well, showing even the Washington-Ball bloodline was not immune for the times.

FRENCH MEDIEVALIST PHILIPPE ARIÈS, IN HIS PIONEERING WORK (MAINLY considered the first) of the history of childhood, *Centuries of Childhood* (1962), argued that during the medieval era of Europe, the notion of what "childhood" was did not exist. In an era of high infant and child mortality, perhaps it did not matter. Horrible as it sounds, an infant was as expendable as any, and the idea that "childhood" was separate from "young adulthood" would have raised a few eyebrows.[12]

Centuries later, the development of childhood became more sophisticated, including clothing that distinguished the youth from adults. This undoubtedly came from philosophers like John Locke, whose 1689 work *An Essay Concerning Human Understanding* argued that there is "little reason to think [newborn children] bring many ideas into the world with them." Children weren't just mini adults, he argued.[13] Yet parents were still uncertain how to treat them. Historian John F. Walzer called it "a period of ambivalence" in which birthdays for infants and children were not celebrated, and, as we see with the simple facts-only notes of the Washington family Bible and various diaries, were mentioned almost in passing.[14] One wouldn't see George celebrate his first birthday with candles, cake, and presents.

But it wasn't out of the question for Mary, or any mother for that matter, to have such affections toward their children. Nancy Shippen Livingston of Pennsylvania wrote several diary entries absolutely gushing over her newborn daughter Peggy. "My sweet Child," she wrote one May day, "my whole soul is wrapp'd up in *you*!" Days later, she wrote, "I allmost [*sic*] devour'd it with kisses . . . I spend so much of my time in caressing & playing with Peggy that I allmost [*sic*] forget I have any thing else to do."[15] It was a sweet private moment between daughter and mother—one that Mary Ball Washington may have very well felt toward her newborn and first son, as well.

AMONG THE DIFFERENCES BETWEEN THE EARLIER CENTURY AND GEORGE'S time was the clothing of children and infants. Baby George—and his brothers and sisters too—wore clothing akin to a dress, with a bodice fastened around the back. For the first three months of his life, George may have been swaddled tightly, but as he matured in the following years, he would have been dressed more like an adult. He also may have worn stays as an infant, a sort of unisex precursor to the corset, designed for proper posture.[16]

MARY, MEANWHILE, COULD NOT PART WITH HER APRON—A PIECE OF CLOTH-ing as familiar then as now, with the same purpose, from tending to the children to cleaning to cooking. Opulent aprons were embroidered with fabrics and rich flowery designs.[17] More intimately, undergarments were unlikely to be seen until the nineteenth century, with the nearest garment to the skin being white linen or cotton or a shift—which also acted as sleepwear. Since Mary was prosperous enough, some shifts (nightgowns, we would call them now), would have been exclusively for the evening. Formally, gowns of silk would have dominated Mary's apparel, with gores and gussets—little inserts to fluff out the dress—abounding.[18]

MARY WAS PROUD OF HER SON. "BEFORE [GEORGE] WAS A MONTH OLD SHE bundled him up, too full of pride to wait another day, and drove away to show him to her favorite cousins, the James Balls of Bewdley," wrote historian Nancy Turner.[19] Located in Lancaster County, that house in Bewdley was burned to the ground in 1917, leaving only one chimney from a one-and-a-half-story building. James Ball, born 1678, was the third son of William Ball, the brother of Mary's father, Joseph. During his nearly eighty-year-long life, the cousin to Mary Ball was married thrice (as was his son, James Jr.).[20]

An apocryphal letter from Augustine made its way to author Moncure Conway through a Washington descendant, saying that, as Conway noted, "He and his wife will make [a certain 'Mr. Jeffries'] a visit on their way to Moratico," where James Ball lived, and that "they will bring with them their 'baby George.'"[21] Historian and Washington expert Douglas Freeman was skeptical of the letter itself due to its second- or thirdhand nature. A trip to James Ball would have occurred, at some point, but taking a long-distance trip with a newborn baby at the edge of the winter season would have been a risk Mary and Augustine would not want to take.[22] Certainly, showing newborns and young children to extended family was not an uncommon practice, and sometimes, in letters recorded, the newborn children were doted on more by the uncles and aunts and grandparents than the actual fathers and mothers![23]

THE CHILDREN OF MARY BALL WASHINGTON, GEORGE INCLUDED, MAY never have been nursed from her. Often, especially in rich families, the infant would have been fed from a wet nurse, either an informal or a legitimate community position, and sometimes by a slave. Though the New England Puritans in the early eighteenth century objected to their use on religious and practical grounds, the upper class of English Virginians continued to employ nurses. They were seen as necessary in a time of deaths during childbirth, as a healthy woman who was lactating may not have been so susceptible to illness as the mother herself, at that time.[24] A woman's time needed consideration as well, and the wife of a plantation owner did not sit idly by while the man did all the work—she was to manage it equally. "Contrary to popular imagination, an early—eighteenth century homemaker of moderately prosperous means spent little time sewing dresses or baking pies." To breastfeed—much less breastfeed more than one child—was time-consuming, and was more likely from someone who was hired to do so.[25]

The wet nurse was also, far less practically, a cultural position; a mother of those times to breastfeed her own children would be a debasement, something that high-class or proper women would never consider. Landon Carter, the son of Virginian colonist Robert "King" Carter, called his own daughter "such a vile, obstinate woman" for her choice to breastfeed.[26]

Advertisements were posted many times in this age, specifically about hiring a wet nurse. One in December 1775 asked for a wet nurse with no children; another, in April 1777, a wet nurse "with a good breast of milk," and if she had child, she could not bring it in.[27]

In the case of Mary, more than likely, the wet nurse she employed was a slave of the plantation, an increasingly common duty through the eighteenth and into the nineteenth century for the South.[28] The irony that they owned other human beings yet trusted them to feed their children may have been lost on them all. It was a simple fact in those days that no one questioned. If a slave wet-nursed George, that was how things were done.

Mary's rapid deliveries—George in 1732, Betty in 1733, Samuel in 1734, John Augustine in 1736, Charles in 1738, and Mildred in 1739—could have pointed to an inability to use breastfeeding as a contraceptive.

GEORGE WASHINGTON WAS CLOSE TO TWO MONTHS OLD IN APRIL OF 1732, when he was baptized.

The length between birth and baptism was unusual, at least for what the Church of England proclaimed. "The Curates of every parish," said the Book of Common Prayer, "shall often admonish the People, that they defer not the Baptism of their Children longer than the first or second Sunday next after their Birth, or other Holy-day falling between." It also warned them that a baptism must be done in a church, unless under special circumstances.

Theologically in all Christian denominations, birth and baptism were to take place as close as possible. To be cleansed of original sin was a precursor to entering Paradise. (If George had died in those two months, he would not have, according to the Church of England, been one with God.) Practically, the Anglican Church functioned as an unofficial arm of the British government. It was the state church, a gatekeeper with the power to give or withhold salvific rituals. Anglican ministers were sworn to obey the king, who, since the English Reformation, had been head of the church. Mary was raised and shaped by Anglicanism. Her son would lead a revolution that would turn many priests into soldiers and split the church along partisan lines into Anglicanism and Episcopalianism. After the Revolution, George would resign from the vestry at Truro Parish, breaking with the former for the latter.

But for now, George was entirely dependent on the benevolence of the Empire.

ON APRIL 5, 1732, THE WASHINGTON FAMILY BIBLE NOTED PLAINLY THAT George, seven weeks old, "was baptized the 5th of April following."

The baptism of George Washington positioned Mary in a life for which she had been prepared since childhood. Ever since she was a young girl, she had been interested in faith. This was no more evident than in her copy of *The Christian Life* by John Scott, published in 1700 in London and sold at the St. Paul's Churchyard. On the front sheet was signed "Mary Ball, 1728," indicating her interest as a young woman.

The Christian Life was a five-volume work by Scott, a High Church Anglican rector of the newly renovated Saint Peter le Poer Church in London and whose life's work, in many ways, was devoted to this tract. The work was a devotional on how to obtain heaven, how to practice the virtues in life, how to practice the faith—all with the ultimate goal of being with God. Mary's interest in this book reveals much about her faith, and her belief in the afterlife. To reach heaven was the ultimate reason for living and minding how one acts on earth. Indeed, when she died in

August 1789, the *Gazette of the United States* ran a notice of her passing, stating that "she conducted herself through this transitory life with virtue, prudence and Christianity."[29] She was a woman of faith unto her death, shown not just in her deeds but in her possessions as well. George Washington Parke Custis, her adopted great-grandson, wrote that she was "always pious, in her latter days her devotions were performed in private. She was in the habit of repairing every day to a secluded spot, formed by rocks and trees near to her dwelling, where, abstracted from the world and worldly things, she communed with her Creator in humiliation and prayer."[30] Mary was known to be extremely devout in theory, if not in practice. Of course, her duty as a mother would be to pass the faith on to her children.

Her oldest would be the first of many children to receive that lesson: all things are done with heaven in mind, they learned. George learned the lesson well. Later in life, he wrote to the Hebrew congregations of Newport, Rhode Island, "May the father of all mercies scatter light, and not darkness, upon our paths, and make us all in our several vocations useful here, and in His own due time and way everlastingly happy."[31] He wrote to Reverend John Rodgers in June of 1783, after the Revolutionary War: "Glorious indeed has been our Contest: glorious, if we consider the Prize for which have contended, and glorious in its Issue."[32] Clearly, this religious upbringing, he and many believed, shaped their call for independence in the very event that defined him and the entire generation.

He may have learned this worldview from Mary. Other books in her collection and interests included lawyer Matthew Hale's *Contemplations, Moral and Divine*, first published in 1676, and James Hervey's *Meditations and Contemplations* from 1750. (Hervey's two volumes are currently at Mount Vernon, signed by Mary Washington and her grandson, Lawrence Lewis.) Hale's own *Contemplations*, also at Mount Vernon, was said to have been read by Mary to her young children.[33] This tradition started with Reverend Edward Charles McGuire, the son-in-law of George's nephew, Robert

Lewis, and longtime pastor of Fredericksburg's Saint George's Church. He noted that the book itself "bears the marks of frequent use, and, it appears, in certain parts, to have engaged particular attention." Among the parts in use was a treatise on humility, a necessary component for both child and woman.[34] Washington Irving wrote that Hale's work, and Mary's teaching of Hale, imposed a sense of "outward action as well as self-government, [which] sank deep into the mind of George." Perhaps, Irving believed, this very work helped form the self-governance not just in character but in philosophy that was so important to the forming of the United States of America decades later. If so, then perhaps Mary's teachings gave a new sense of building the nation beyond what anyone possibly imagined—now and certainly back in the 1730s. "Let those who wish to know the moral foundation of his character consult its pages," Irving said.[35] It may have been as much a founding document of this nation as any other.

These works were all designed to spiritually enlighten one's life on earth in preparation for the next, for deeply devoted men and women of faith, by means, as Mary Thompson wrote, of using "examples from nature to teach about the nature of God." Many of these authors were open to using the "reasonableness of Christianity" to defend the faith in an Enlightenment setting, rejecting the mystical for the empirical.[36]

GEORGE WAS WRAPPED IN A SPECIAL CEREMONIAL ROBE DURING THE SACRAment, befitting of such a solemn event. (And it exists to this day. For a short period in 2017 and 2018, the Smithsonian Institution's National Museum of American History, for its exhibit Religion in Early America, publicly displayed what is thought to be George Washington's baptismal robe and blanket.)

In 1923, the famed explorer Robert Shackleton wrote that "there is a child's christening robe, of white silk brocade, a robe so delicate and attractive it would draw attention even if it were not associated with any known individual."[37] The 1.5-by-1.5-square-foot silk cloth had that "attractive" look, with a pinkish red silk interior and white exterior, embroidered with decoration. It, along with a plethora of other "relics," was transferred to the Smithsonian from the United States Patent Office. They were purchased in 1878 from George Washington Lewis's family for $12,000.[38]

The colors passed into history, some say, as an inspiration for the Stars and Stripes, designed decades later, signifying a further Christian origin of the nation. It's a preposterous legend that is a mere coincidence, at best, but Helen Richardson's article in *The Churchman* in February 1904 purported an addition to the myth of the Father of His Country. His inspiration, she said, was so great, that his baptismal robes were the colors of the nation.[39]

ALL OF THIS, FROM THE DOCUMENTATION TO THE ACTUAL MATERIAL, CONtradicts one particular legend of George's baptism. The grandchildren of the Reverend John Gano, a Revolutionary Army chaplain and minister of First Baptist Church in New York, first reported at the end of the nineteenth century that their grandfather honored George Washington's request to be baptized by immersion during the winter at Valley Forge at the end of 1777. "I have been investigating the Scripture," George purportedly said, according to a *Time* magazine article in 1932, "and I believe immersion to be baptism taught in the Word of God, and I demand it at your hands. I do not wish any parade made or the army called out, but simply a quiet demonstration of the ordinance."[40]

Of course, Gano himself did not mention this event in his autobiography, nor did any witnesses come forward. Facts changed as decades went on. One painting depicted the immersion in the Potomac—nowhere near Valley Forge. It did not happen. George was a lifelong Episcopalian and would have seen his infant baptism—an event that is one and done, never repeated—as valid.

A SCANT THREE YEARS AFTER GEORGE WASHINGTON WAS BORN, HIS NAMEsake, the guardian of his mother in her youth and in many ways a second father to her, George Eskridge of Westmoreland County, died at Sandy Point. He was seventy-five years old by the time of his death on November 25, 1735.

We can only imagine Mary's grief upon hearing of the death of her guardian, a man who influenced her so. He died at an old age, breaking the chain of young deaths in Mary's life.

Today, in Fredericksburg, stands a mighty oak tree near the intersection of Washington Avenue and Pitt Street. It was moved there from Sandy Point on April 29, 1937. Near the towering memorial to Mary Washington herself and the cemetery of the Grove family of Kenmore, a plaque reads:

COL. GEORGE ESKRIDGE MEMORIAL TREE

APRIL 29, 1937

MAY THIS OAK TREE FROM "SANDY POINT" WESTMORELAND CO. VIRGINIA, HOME OF COL. GEORGE ESKRIDGE, WHO WAS GUARDIAN FOR MARY BALL, SHELTER HER LAST RESTING PLACE, AS SHE IN HER EARLY CHILDHOOD WAS SHELTERED AND PROTECTED BY HER BELOVED GUARDIAN. AS DESCENDANTS OF OUR ILLUSTRIOUS ANCESTOR, WE DEDICATE THIS TREE TO THE MEMORY OF OUR COUNTRIES NOBLEST MOTHER AND HER GUARDIAN, COL. GEORGE ESKRIDGE.

MRS. ELISE TOWSON COELE

SPONSOR

THESE FIRST YEARS OF INFANT GEORGE'S LIFE AT POPES CREEK SAW THE growth of the Washington family. Augustine and Mary Washington gave birth to Elizabeth—"Betty," as her family, friends, and history have known her—on June 20, 1733, only a year after George's own birth. She was born about six in the morning. Then, at three a.m. in November of 1734, Mary gave birth to her second son, Samuel. At two in the morning in January of 1736 came John Augustine Washington, and in May of 1738 came Charles.[41]

NO LATER THAN JANUARY OF 1735, THE WASHINGTON FAMILY—AUGUSTINE, Mary, George, Betty, and Samuel, but without the three children from Augustine's first marriage—moved north over seventy miles. That was about a day's walk on the west side of the Potomac River to Augustine's 2,500-acre tract of land of Little Hunting Creek— the future plantation of Mount Vernon. (Augustine inherited this land in 1726, after his sister agreed to sell him the estate.) A letter and enclosed memorandum from Hannah Fairfax Washington to her cousin George indicated that they moved in 1734,[42] but either way, George was less than three years old, barely aware of his surroundings in Westmoreland County, when they packed up. From that time and for the next 123 years, this patch of land was owned, occupied, and operated by a Washington and Washington descendants before it was sold to the Mount Vernon Ladies' Association in 1858.

Why the Washingtons moved north to the isolated region is up for debate. Dr. Philip Levy, a professor of history at the University of South Florida, called the move "curious," as Popes Creek was the "epicenter of the Washington empire for three generations."[43] A variety of valid and invalid reasons exist. Many biographies falsely claimed that Popes Creek Plantation, also known as Wakefield, the birthplace of George Washington, burned to the ground in 1735. In reality, it burned to the ground on Christmas Day, 1779.

Most likely, the move placed Augustine closer to his work at the Principio Company's Iron Works. It also placed them closer to Mary's 600-acre inheritance nearby, which cemented his landholdings in Prince William County. Luke Pecoraro, then director of archaeology at Mount Vernon, believes this was the driving force behind the move. "In the end, follow where the money goes," he said.

Popes Creek was primarily a tobacco plantation. With the iron furnace nearby, "was Little Hunting Creek a chance to expand on that?" Pecoraro continued, "Prince William County is attractive to Augustine. It is the frontier. If you have 2,500 acres up here already, what can you do with it? It's attractive for development."[44] Indeed, during these years, Augustine was seen in business ventures like no other period in his life. He traveled to and from England in the best interest of the Washington family. He became a one-twelfth owner of the entire company. This was signed in a contract dated April 15, 1737, between Augustine Washington of Prince William County and William Chetwynd of Beddington in the County of Surrey, England, among other owners.[45]

There at Little Hunting Creek Plantation the Washingtons lived until late 1738 when they moved back south, to a plantation in Fredericksburg.

THESE FOUR YEARS OF THE WASHINGTON FAMILY'S LIFE, THOUGH FEW AND far between in facts, have been relatively glossed over.

The first historian to give us insight into this lost period was eighteenth-century Virginian Reverend Philip Slaughter of Emmanuel Church in Culpeper County, whose father was a captain in the 11th Virginia Regiment during the American Revolution. Slaughter was the preeminent historian of the Truro Parish region, which made him an invaluable resource for information about the Washingtons. He discovered the records of vestrymen of the newly installed Truro Parish. Author Moncure Conway spoke and wrote to Reverend Slaughter before the reverend's passing in 1890.

In a letter to Conway dated July 24, 1889, Slaughter detailed his findings, giving a time line of events from October 1737 to October 1739, when Augustine's "name does not appear again."[46]

On November 18 (or 28), 1735, Augustine Washington was elected and sworn in as a vestryman of the parish. His and Mary's male ancestors were vestrymen of various parishes in Virginia, and their ancestors continued to be vestrymen, so it isn't exactly a radical characteristic of George's father. Yet it does show that they became acquainted with others near Little Hunting Creek, enough to be elected to the body of the parish. In August of the following year, 1736, Augustine recommended Charles Green to become minister of the parish. It was approved, pending word from the bishop of London. Reverend Green held the position of rector of Pohick Church for the next twenty-seven years, until 1764.[47] Augustine's influence was strong enough—and presumably his demeanor was familiar enough—for the other vestrymen of this brand-new church to have him select its rector.

ACROSS THE OCEAN, IN ENGLAND, THREE THOUSAND MILES AWAY, A BABY boy was born on June 4, 1738, at Norfolk House in St. James's Square, Westminster. He was named George William Frederick: George after his grandfather, and Frederick after his father. A little over a decade earlier, his grandfather was crowned as king of Great Britain, becoming George II.[48]

The little boy grew up under this reign and would succeed his grandfather after the death of his father, the Prince of Wales. He would become King George III of Great Britain.

While that little boy named George, born in England, would grow up to become king, the little boy named George, born in Virginia, would grow up to become president.

OTHER EVENTS OCCUPIED AUGUSTINE'S LIFE DURING THESE YEARS. HE had runaway slaves and indentured servants. There was no better place to advertise and offer a reward for a runaway than, of course, the *Virginia Gazette*. So said the bulletin, from page 4 on June 9, 1738:

> *Ran away from Capt. McCarty's Plantation, on Popes Creek, in Westmoreland County, a Servant Man belonging to me the Subscriber, in Prince William County; his Christian Name is John, but Sirname [sic] forgot, is pretty tall, a Bricklayer by Trade, and is a Kentishman; he came into Patowmack, in the Forward, Capt. Major, last Year; is suppos'd to have the Figure of our Savoir mark'd with Gunpowder on one his Arms. He went away about the 20th of April last, in Company with three other Servants, viz. Richard Martin, is a middle siz'd Man, fresh colour'd, about 22 years of Age, and is a Sailor; had on a blew [sic] Jacket. Richard Kibble, is a middle siz'd young Fellow, has*

several Marks made with Gunpowder on his Arms, but partic-
ularly one on his Breast, being the Figures of a Woman and a
Cherry-Tree, and is a Carpenter by Trade; he wore a blew grey
Coat with a large Cape, a Snuff-colour'd Cloth Wastecoat [sic],
and Buckskin Breeches. Edward Ormsby, is a small thin Fellow,
of swarthy Complexion, and is a Taylor by Trade; has a Hesita-
tion or Stammering in his Speech, and being an Irishman, has
a good deal of the Brogue. They went away from Capt. Aylett's
Landing, on Patowmack, in a small Boat, and are suppos'd to
be gone toward the Eastern-shore, or North-Carolina. Whoever
will secure this said Bricklayer, so that he may be had again,
shall have Five Pounds Reward, besides what the Law allows,
paid by
    Augustine Washington
    N.B. It is not doubted but the Owners of the other Servants
will give the same Reward for each of theirs.[49]

The advertisement appeared three times; it is unknown if John
the Kentishman was ever captured, but the omission of the nota
bene at the end in the latter two letters indicated that he was not,
while the others were.[50]

IF A LACK OF CONCRETE INFORMATION ABOUT THESE EARLY YEARS OF
George's life, and how he was raised, went unnoticed, then historians could take solace at every little piece or hint of information
of how he may have been raised, even indirectly. Buckner Stith,
a childhood friend who was ten years older, described his relationship with Lawrence Washington, George's older half brother,
in a letter decades later. He assisted the Washington family in
1738, when he was sixteen years old. "I am the same Man who
marched with him [George] and old Laurence from Chotanck to
Fredericksburg, how Laurence and him laughed at me for holding
the wine glass in the full hand, but as I was five Years older than
either of them, I thought I might hold the wine glass as I pleased;

that we lost a Horse or two in the Trip, and were obliged to walk honestly in turn clear to Chotank again." He described George at that time as "a sound looking, modest, large boned young Man."[51] A six-year-old George laughing with his older brother at a bumbling friend.

The first six years of George's life are lacking in detailed facts beyond a date here or there. Accounts of Mary's early motherhood are nonexistent. It is not known in detail how she definitely raised or treated her children, whether she went with or against colonial norms. But George's early life, Ron Chernow wrote, "was a roving and unsettled one," moving to and from plantations in Virginia.[52] Certainly, from Popes Creek to Little Hunting Creek, there was a major shift, upending a little boy's life for something unfamiliar. What friends George made in Westmoreland County had no reason to visit Prince William County. He had his immediate family and slaves' families to play with here. "City boys and girls," wrote Kate Wiggin and Nora Smith in their children's book in 1891, "might think, perhaps, that little George Washington was very lonely on the great plantation, with no neighbor-boys to play with; but you must remember that the horses and cattle and sheep and dogs on a farm make the dearest of playmates, and that there are all kinds of pleasant things to do in the country that city boys know nothing about."[53]

There's a romantic conception here of being alone in a quiet landscape, with just the animals and crops to look over. The decades before the Industrial Revolution offered a silent world free of the bustle of a mechanized future.

Here George met his half brothers, Lawrence and Augustine Jr., who had been away at Appleby School in England, expanding the knowledge of the family to Augustine's own family.[54]

The Washingtons' closest neighbors were the Fairfax family in Belvoir. This would prove somewhat beneficial for the children— Lawrence would marry Anne Fairfax, daughter of Colonel William Fairfax, who himself would see great promise in George in the following years.[55] At the other end of the spectrum, George was deeply smitten with Sally Fairfax, who, by all accounts, was beautiful and charming.

AUGUSTINE SPENT A GREAT DEAL OF TIME IN ENGLAND FOR BUSINESS. THIS was not a safe passage, either. Disease was rampant. The *Virginia Gazette* often advertised and posted bulletins of Augustine's return, such as on July 22, 1737, which read, "We here from *Potowmack*, That a Ship is lately arriv'd there, from London, with Convicts. Capt. Augustine Washington, and Capt. Hugh French, took their Passage in her; the Former is arriv'd in Health; but the Latter dy'd at Sea, and 'tis said of the Goal Distemper, which he got on Board."[56] Whether through luck or good hygiene, Augustine survived numerous trips in cramped, dangerous conditions that could have easily led him to death, all before George would have a clear memory of his father. It was with luck that George did not become fatherless at such a young age, as Augustine and Mary themselves had.

Little George became oriented with the customs of the Virginian plantation as he grew. He learned to walk, talk, and read with picture books and poems and hornbooks, like his mother before him. He learned social cues from those around him. He learned a mixture of parental obedience and self-reliance, a delicate balance. "Not only did young children experience considerable freedom of movement on the plantation, but they also lived under few parental restraints to their conduct. Parents and kin, at least in middle- and upper-class families, apparently made little effort to stifle childhood willfulness and self-assertion." Diaries from parents congratulating their children for being so autonomous were seen throughout the eighteenth-century Virginia and Maryland.[57]

But from the relative freedom of the plantation, George was about to move into the heart of a royalist city. Not only was the town itself named after the successor to the throne, but nearly every street was named after a royal person or place or association.

Sophia Street was named after King George's sister; Hanover Street after the House of Hanover; and so on.

At the age of six, George was to move to a new location. This time, back south, about forty miles.

To Fredericksburg, Virginia.

Chapter 6

Fredericksburg

LIFE ON HOME FARM, FREDERICKSBURG

1738–1743

---

*"The homes of her childhood and her
early womanhood lapsed away."*

T he move to Fredericksburg was only the first step into a
broader world for young George Washington. It was also
the site for many of the myths which have sprung up
around his childhood. For Mary, it was to be home for the rest of
her life. She took one more step into the world and stopped. She
would raise her children here, be widowed here, and die here.

A quarter-page ad in the reliable *Virginia Gazette* appeared on
April 21, 1738. It may have caught Augustine Washington's eye:

*To be Sold, for Cash, on the 25th of October next, by way of
Auction, to the highest bidder, several Tracts of Land belonging
to the Estate of William Strother, late of King George County,
Gent. deceas'd, pursuant to his Will, wiz.*

*One Tract, containing 100 Acres, lying about 2 miles below
the Falls of Rappahannock, close on the River Side, with a very
handsome Dwelling house, 3 store houses, several other convenient Out-houses, and a Ferry belonging to it, being the Place
where Mr. Strother liv'd; is a beautiful Situation and very commodious for Trade.*

One other Tract, of 160 Acres of very good Land, adjoining
thereto, the Plantation, Houses, Fences, &c., in good Order.

About 1,600 acres were also for sale in Prince William, along
with twenty slaves.[1] William Strother, the previous owner, had
died several years earlier, in late 1732 or early 1733; his will, made
out in November of 1732, equally divided his main estate among
his six children (all daughters) and his wife, Margaret—all other
lands, which included the future Ferry Farm, were to be sold if
Margaret so desired. As she had taken another husband, who had
an estate of his own, she decided to sell the extra land.

As with some ads of the era, there was a hint of exaggeration.
The estate did not operate a ferry, though one disembarked on
the land after crossing the Rappahannock. The "very handsome
Dwelling house" may not have earned that description. One man,
as recounted to author Moncure Conway some decades later, recalled that his father described it as simply "a plain wooden structure of moderate size, and painted a dark red color."[2] "Moderate"
is in the eye of the beholder. The house by its very nature, though,
had to be large enough to accommodate Augustine and Mary and
their children. An inventory listing of all possessions in each room
years later also noted that there was a hall, two back rooms, a parlor, a passage, a hall chamber, a parlor chamber, a dairy, a closet,
a store house, and a kitchen, each with substantial furniture and
possessions, indicating large sizes.[3]

Augustine learned of the sale, and on November 2, 1738, bought
about 280 acres of the land for 317 pounds. It was an all-around
good move. It was closer to the Principio Iron Works than Little Hunting Creek, and closer to familiar land of Popes Creek.[4]
Mary's half sister, Hannah, lived in Stafford County, perhaps providing a well-known anchor for moving back south. She "had to
some extent come home."[5] She was only thirty miles from Westmoreland County, and Lancaster was also relatively nearby.

Here, Mary would have a sense of "individuality." Author Nancy

Turner wrote, "The river farm became a part of the very fabric of this woman's life. . . . The different features of the simple surroundings on take special associations. . . . The homes of her childhood and her early womanhood lapsed away into little more than pleasant dreams."[6] Mary lived on this farm until 1772, when she moved to nearby Fredericksburg, where she spent the rest of her days, and thus her "home" of Little Hunting Creek and Popes Creek and the childhood residences of Epping Forest, Sandy Point, and Bonum's Creek faded into the mist, as Ferry Farm superseded all.

The "Rose of Epping Forest" had moved on.

ORIGINALLY KNOWN AS HOME FARM TO THE WASHINGTON FAMILY, THIS childhood home of George was settled closely to the southern Virginian hub of Fredericksburg, in then King George County. He inherited it upon the death of his father yet continued to call it "my mothers [*sic*]" farm many years into his adulthood.[7] It became known as "Ferry Farm" a century later due to the massive ferries going back and forth across the Rappahannock River. It now bears the distinguished title of George Washington's Boyhood Home.

The view offered was unparalleled compared to anything the Washington family had possessed. An advertisement for the land for sale appeared in an issue of the *Virginia Gazette* on November 5, 1772, decades after the initial move, which described it as "a tract of six hundred acres . . . one of the most agreeable situations for a house that is to be found upon the whole river, having a clear and distinct view of almost every house in the said town, and every vessel that passes to and from it."[8]

William Byrd wrote a description of the town at the time, around the initial move, in 1732: "Sloops may come and lie close to the wharf, within thirty yards of the public warehouse which is built in the figure of a cross. Just by the wharf is a quarry of white stone that is very soft in the ground, and hardens in the air. . . . Though this be a commodious and beautiful situation for a town, with the advantages of a navigable river, and wholesome air, yet the inhabitants are very few."⁹ The region was not flat; instead, it had hills and slopes in every direction as merchants went to and from the Rappahannock. The Rappahannock itself, appropriate for its mighty-sounding name, moved forward past Fredericks-burg for miles, one of the most important trade routes for the region. The headwaters originated nearly two hundred miles west.

THE HOUSE ITSELF WAS MODEST FOR ITS ACREAGE AND PRESTIGE, YET EX-uded gentry at every turn. The twenty-plus slaves made sure of that; without the slaves, the Washingtons would have been sty-mied, stuck with an unmanageable plantation. The back hall room, on the south side, adjacent to the back porch, had a four-poster bed, the most expensive bed in the house. Mary and Augustine slept in this room. Here, too, tea for the well-heeled ladies was often served, and Augustine managed the house from here. Half a decade later, it was in this same small room that Augustine fell ill and died.

The hall room itself, north from the back hall room, contained a small but fancy fireplace, and it was the largest and nicest room, where the family and guests ate. Through the windows, they could see the hill sloping down to the Rappahannock River, a stone's throw away.

The parlor was the social space. There was just one window, but a huge fireplace, over five feet tall. Here, daughter Betty would have been courted by Fielding Lewis; and the boys and Augustine, and guests as well, would have played typical card games like whist. The discovery of ceramic wig curlers—another status symbol—showed that the boys would've also been dressed and presented here. The oldest furniture was placed here, creating an illusion of status without depleting the finances.

Some years later, around 1740, Augustine built an additional room, the back room. A small study, with an equally small fire-place, the room has left historians and archaeologists scratching their heads as to its purpose. It was possibly multipurpose, where the family hosted high-profile guests or even children who wished to stay the night.

Upstairs were the parlor and hall room chambers, where the family slept and stored lumber and other materials.

The Washingtons' garden was in the lower hills to the south; Orinoco tobacco (instead of the more valuable and better sweet-scented tobacco) and crops were sustainable, though not great.

All in all, size was important, but it was how the family used their home and grounds which counted most.

And, after all the years of moving back and forth, this house would be called home.

IN THE SHORT FIVE YEARS BETWEEN THE MOVE FROM PRINCE WILLIAM County and Augustine Washington's death in 1743 were the legends of the famous cherry tree incident and other tales, small and large, little known and well known. What Popes Creek was to young George's birth, the Home Farm was to his childhood. Home Farm itself shaped the young boy, leaving the rural northern Virginian region for a farm near an urban trade town.

The actual move southward occurred in November of 1738; by December, they were settled in. By that time, Augustine had acquired an additional 380 acres adjoining the land from the original purchase.[10] Their home at Ferry Farm was situated just southeast of modern downtown Fredericksburg.

Located in newly established Spotsylvania County, and barely even a town at the point of the Washingtons' move southward, Fredericksburg was established in 1728 as a settlement by an act of the Virginia General Assembly (as was the town of Falmouth in King George County).

IN THE WORDS OF HISTORIAN PAUL WILSTACH, FREDERICKSBURG WAS "NOT a bad place to keep in touch with the world." It was a growing trade route, with Williamsburg to the south and the Rappahannock and Potomac to the east. King's Highway ran through into the northern colonies. It may have been sparse, but it had all the markings of a booming frontier.[11] The discovery of iron ore on the Rappahannock shores sometime after certainly made a good impression on hopeful prospectors.

Certainly, Mary and company would have been interested in the imports and trade and quickly acquainted themselves with the area; in 1746, Captain John Kerr in his ship *Restoration* arrived from London with everything from Dutch metals to sweet almonds, medicine, and barley, all for sale.[12] Apple and cherry orchards in and around the city would spring up soon enough, horse races and fairs too, all making this trade route into something, something more permanent—something like home.

Governor Alexander Spotswood moved the county seat in 1732 to this new tiny budding town with a population of six. By 1782, the county had grown to 5,500 people. This included immigrants from Scotland, and such names as Fielding Lewis, Henry Willis (the second husband of Mildred, Augustine's sister), and William Byrd. Each of these in the nearby neighborhoods saw the "frontier" of Virginia as a viable and potentially profitable land.[13]

And so did the Washingtons. The location of their new home was so opportune: the river, King's Highway, a growing town, all within a very short distance. It meant the Washingtons' lives were "quite literally at the crossroads of local life and commerce."[14]

GEORGE WASHINGTON GREW UP HERE. HERE, ALSO, STORIES OF HIS YOUTH developed. Parson Weems, the origin of many of these fanciful tales, recounts the most famous of all these stories, as told to him by an anonymous "aged lady, who was a distant relative" in 1833: "One day," he wrote, "in the garden, where he often amused himself hacking his mother's pea-sticks, he unluckily tried the edge of his hatchet on the body of a beautiful young English cherry-tree, which he barked so terribly, that I don't believe the tree ever got the better of it." Augustine Washington was flabbergasted, confused, and wanting to know who did it. "Presently, George and his hatchet made their appearance. 'George,' said his father, 'do you know who killed that beautiful little cherry tree yonder in the garden?' . . . He bravely cried out, 'I can't tell a lie, Pa; you know I can't tell a lie. I did cut it with my hatchet.' "[15]

And so the legend was born.

The morals of honesty, bravery in the face of punishment, and taking responsibility for mistakes presented a timeless lesson for children of all ages and all generations. Perhaps George learned this very lesson at some point. The little six-year-old boy would have needed to learn this from his father and mother, of course.

There are, again, some grains of truths in legends. Perhaps George did not cut down a cherry tree with his hatchet. But the story possibly came from somewhere, and some influence and enjoyment of cherry trees for the Washingtons were evident. This came from a 1760s punch bowl, owned by Mary Washington. It depicted red bundles of cherries (a rare and therefore pricey design), with stems and leaves. But more important, it was cracked. As Dr. Philip Levy, whose archaeological team discovered this artifact, stated, a cracked bowl was "usually a death sentence for table ceramics." This one, however, "was special enough to Mary that she paid a craftsman to repair the cracks with a special glue paste," possibly made of cheese. "It was not as good a drinks vessel as when it was new, but it could still hold pride of place on a shelf or a mantle."

Levy believed that, artifact of cherries or not, this did not point to Weems's legend as fact. Some historians may have very well said so, without missing a beat. Some may have even placed a whole mythology as to why Mary bought or received this cherry bowl—her son's loyalty was so important to her, she must have a sort of relic in honor of it! Some may have made mythical paintings of the bowl's use, perhaps even its purchase or placement in the Ferry Farm plantation. But the bowl "crystallized" the legend: "If George Washington did indeed chop down the Cherry Tree," Levy said, "as generations of Americans have believed, this is where it happened."[16]

Another well-known tale told of a little mischief and white lies from Augustine. In order, according to Weems, "to startle George

into a lively sense of his Maker," the father went into the garden of Ferry Farm, dug out the soil to form big letters spelling
George's name, and filled them with cabbage seeds. Some weeks
later, George, ecstatic when he saw the seeds fully grown, called
over Augustine. "O Pa! come here! come here!" A little child,
impatient to show some great wonder—for that is exactly what
George must have been at that time—ferociously tugged at his
arm and brought him to the garden, where, "in all the freshness
of newly sprung plants," the name GEORGE WASHINGTON
was embedded.

Little George was awestruck. But he was smart, and deduced
that surely it couldn't have just sprung there. Someone—his father, George concluded—must have planted it there. At that point,
Augustine admitted, "I indeed did it; but not to scare you, my son;
but to learn you a great thing which I wish you to understand. I
want, my son, to introduce you to your true Father." For just as
George did not see Augustine plant the seeds, George knew, he
could not see God, but knew of His existence. It wasn't "chance"
that made the name appear, it was created. So too, Augustine said,
were "all those millions and millions of things that are now so
exactly fitted to his good!"[17]

It was a religious lesson of the invisibility yet omnipresence of
the Creator, who gave George everything he knows or knew or
will know, everything he has, had, or will have. Augustine was
an invisible creator of the garden, and so God was the invisible
Creator of all, as the Washington family believed.

ULTIMATELY, THE LEGEND OF THE CHERRY TREE AND THE GARDEN WERE
tales of a father-son relationship. Parson Weems may have very
well made these up as a lesson to teach fathers how to raise their
sons. They added a level of authenticity to the relationship. Augustine was an honorable fellow, having been elected a trustee of
Fredericksburg several times. He was a "friendly man," according
to one English physician.[18] Families in the region were more or
less hands-off with young children, but the father of the household

increasingly gave lessons, and expanded the child's experience of what was to be expected of him when he matured into adulthood. This included not only George's dressing like an adult man, moving from the unisex dresses of his infancy, but also lessons in the everyday life of running a plantation. His father was his idol, a role model for the perfect man to the little boy. Six- to ten-year-old George may very well have attended business meetings with his father, learning the ins and outs of Virginian plantation life.[19]

Where Mary fit at this time was unknown. As a woman, she probably had "considerably less influence than" Augustine over her sons. A few decades after her children matured, in the early Federalist period, children of Virginia and Maryland needed their mothers' approval, especially in school.[20] But that was almost a civilization ahead.

Her horseback-riding expertise and interest in equestrianism may have helped George with his own. He would have been taught by both Augustine and Mary, but Mary's own interest may have made it a mother-son activity. George may have been a messenger for the Washingtons when both were busy, providing ample opportunity for the young boy to learn of the region and neighbors.[21]

These skills obviously came in handy in George's later life. Indeed, sixty years later, former president Thomas Jefferson, writing from his retirement at his home in Monticello, referred to George Washington as "the best horseman of his age, and the most graceful figure that could be seen on horseback."[22]

Mary's interest in horses continued well past teaching age for her children. Recreationally, horses were used for entertainment in racing, especially breeds imported from England, and the Washingtons and their neighbors owned several dozen, ranging in name from Aristotle and Bolton to Vampire and Sober John. Fredericksburg itself was a site of a horse race in October 1774.[23] Ads appeared frequently in the *Virginia Gazette* on a range of equestrian topics—collars and clothes for sale, trading, lost and found, races. But they were just as much property as hobby. In 1747 in the *Gazette*, Mary Washington posted a notice of a runaway or stolen horse (she was not sure which), which disappeared over two weeks earlier on the night of April 30. It was described as "a large Black Horse, branded on the far Shoulder LW; has a white Spot on each Flank, and likewise on each Side of his Breast; with a long Mane, most Part of which hangs on his right Side." Whoever brought it to her—the notice did not appear again, so presumably someone did—earned up to 1 pistole (a gold coin, or doubloon), worth a little less than but nearly 1 pound.[24]

ON CHRISTMAS EVE, 1740, THE WASHINGTON FAMILY'S LIFE WAS UPENDED. IT was supposed to be the eve of celebration, one of the most joyous times of the year. Instead, on a cold day in late December, their house literally burned down. Or, at least, so said oral tradition. George Washington noted to David Humphreys in his authorized biography that his "father's house burned," though he did not specify where, when, or how, leaving historians to wonder if he meant his Ferry Farm or another dwelling of Augustine's. A neighbor to the Ferry Farm Washingtons, Robert Douglas, an immigrant from Scotland who worked as a clerk in Fredericksburg, wrote in 1795 to then former president Washington that "on a Christmass [sic] Eve, his great house was burned down & that he was Obliged with his good family to go and live in the Kitchen."[25]

This story—even with a letter from a witness—came under historical scrutiny, as Philip Levy and his archaeological team in 2008 could find no evidence of a massive fire, as no distinct structure or foundation existed pre-fire and post-fire. "Repair, renovation, and abandonment were what we saw. But no fire."[26] Instead, what they later discovered was not a house-destroying fire, but only a fire in a single room at the south end of the house, sometime before 1750. It was in the back hall room, which contained a small fireplace. Possibly, burning logs fell away, causing the fire to spread to the room. The burned pieces of plaster discovered in a small cellar indicated that. What was destroyed structurally was quickly repaired, almost seamlessly.[27] A sort of middle-ground narrative was born: It wasn't an inferno, nor was it nothing. It was enough to be noted in one or two historical letters in passing, and that was it.

A fire breaking out, no matter how small, was scary enough—to an eight-year-old George, it must have looked and sounded terrifying. The adults and servants and slaves frantically getting buckets of water to stop the spread; living in the separated kitchen as repairs commenced in the cold would have only added stress to the unfamiliar situation.

But it was repaired, and life went on. House fires at the time were quite common.

AS FORTUNATE AS THEY WERE IN DECEMBER NOT TO LOSE THE ENTIRE plantation house, the family was not as fortunate a few months earlier. On October 23, 1740, Mary's sixth and final child, whom they named Mildred after either her father's sister or her father's mother, died at the approximate age of sixteen months. It is unknown what killed her. Mildred was born less than a year after their initial move to Home Farm. But her death was a wake-up call—a grave memento mori for the Washingtons, who'd been relatively lucky in terms of infant or child mortality. Mildred was not the only Washington, though, who died; Augustine's fourth child, his last with his first wife, died five years earlier. Jane Washington was twelve. He was barely three years old, and in George's young life, death would not be uncommon.

Now it was Mildred who died. George, at eight years old, had a greater understanding of death as more than simply "not coming back" and something incomprehensible.

It was bound to happen at some point. Again, infant mortality was almost a guarantee in every family, and luckily Mary's last was her only. Little Mildred was born at Ferry Farm and died there, offering bookends to the otherwise routine lifestyle of the Washingtons, and for Mary in particular.

It's worth noting that in September of 1771, before selling the land, George Washington conducted a survey of Ferry Farm, opening it at "the little gate by the tombstone." A curious but telling starting point of the land survey, this little tombstone was significant to George so many decades later. Baby Mildred wasn't simply a lost sibling.[28]

Death in the biographies of Mary were almost glossed over, as if it was a distraction to the grand character of Mary and her relationship to her children. The hagiographical *Story of Mary Washington* summarized the birth and death of this child in two sentences, before skipping years ahead to the death of Augustine.[29] Future president Woodrow Wilson's biography of George Washington in 1896 did not mention any Mildred or youngest daughter.

Mary must have been beside herself. It was a great test of her strength of will and fortitude. Death was an "inescapable phenomenon," but it also often served to unite families and communities.[30] The commonality of losing a child offered a sort of solace and solidarity for the grieving mother. Most mothers experienced this very same dread and fear and heartbreak—sometimes more than once.

"God takes away . . . God's will be done," said William Byrd when his infant son, Parke, died in 1710. Perhaps the deeply religious Mary thought the same—that Mildred's death was God's will. But an inseparable and deeply unique connection between mother and daughter exists, which Byrd could not, or would not, understand; however, he did note that there was a difference in his wife, the mother of the dead Parke, as she was "very much afflicted [with] several fits of tears." Weeks went by until she began to compose herself.[31]

THE MARRIAGE OF MARY AND AUGUSTINE WASHINGTON LASTED TWELVE years. This decade-plus required delicate coordination between husband and wife, and mother and father, to manage the plantations, which had grown considerably, and the nine children of both of Augustine's marriages. They had come and gone from Popes Creek to Little Hunting Creek to Home Farm, and then everything changed. Augustine, the father of George Washington, took ill after exposure to the frigid elements. Benson Lossing described the scene:

One day early in April, 1743, Mr. Washington rode several hours in a cold rain storm. He became drenched and chilled. Before midnight he was tortured with terrible pains, for his exposure had brought on a fierce attack of hereditary gout. The next day he was burned with fever. His malady ran its course rapidly....[32]

The patriarch of the Washington family was dying. This was no small event, for either the family or the slaves and servants. "In whispered voices, or hidden discussions, all would have been inquiring about what was said of their name or those of their loved ones. Who would go where? Who would remain together and who would have to leave. . . . The death of a master was every bit as great a tragedy in the yard as in the home—perhaps even a greater one."[33] The livelihood—and lives—of the slaves depended entirely on what was written in Augustine's will.

Augustine knew his end was coming. His will was written on April 11, 1743. He died a day later, at the age of forty-nine. His children ranged in age from five to twenty-five; George was eleven, and later would have only a passing memory of his father. What mentions George made of Augustine later in life were mainly legal, having to do with his lands and will.

His last will and testament, in which he noted he was "sick and weak but of perfect and disposing sence [*sic*] and memory," was lengthy and that of a man of wealth who owned much land. It began with a simple "In the name of God, Amen." Each of his sons from both marriages received plantations of their own. He "left all his children in a state of comparative independence."[34] Lawrence Washington received 2,500 acres of land near Hunting Creek. It was no surprise this move was seen as making Lawrence, the eldest son, the head of the Washington household, as Little Hunting Creek was seen in many ways as the primus inter pares, the first among equals, of all his estates. To Augustine Jr., he gave part of an estate at Bridges Creek in Westmoreland, including cattle and slaves (specifically mentioning a slave named Frank). To George, he gave the estate of Ferry Farm, along with half of the estate at Deeps Run and ten slaves. Samuel Washington received 600 acres at Chotank, Washington's birthplace, and his other sons received other lands and property.

Mary received five slaves, named Ned, Jack, Bob, Sue, and Lucy; the crops at Bridges Creek, Chotank, and the Rappahannock estates; and the rest of her husband's property that was not mentioned was to be divided among her and her children. It was further written, "It is my will and desire that my said four sons [*sic*] Estates may be kept in my wife's hands until they respectively attain the age of twenty one years, in case my said wife continues so long unmarried."[35]

This last part, knowing what we know now centuries later, assumes much, for a simple reason.

Mary was not like other women of her time. If other events in her life didn't say as much, that fact that she remained a widow for the remainder of her life after her husband's death did.

THE DAY AFTER AUGUSTINE'S DEATH, ON APRIL 13, 1743, ABOUT FIFTY MILES to the southwest in Albemarle County, a boy was born. He was the third of ten children; his father, Peter, was a surveyor and plantation owner.

Much later, this boy would grow up to become the author of the Declaration of Independence, and later serve as third president of the United States.

As the Washingtons were grieving at the death of Augustine, so a nearby family was celebrating at the birth of a boy named Thomas Jefferson.

SEVERAL MONTHS AFTER HIS DEATH, IN JULY OF 1743, AUGUSTINE WASHINGton's entire estate was inventoried. The survey tallied and valued a massive list of possessions. There were twenty slaves at Home Farm, each named, ranging in worth from 35 pounds—his name was Bob—to 1 shilling—Jo, possibly a newborn or child, or, conversely, an elder. An additional seven slaves were stationed on other property, called the "Quarters," below Fredericksburg, possibly the living quarters of some slaves. Livestock, including six oxen, nine cows, six calves, twenty-one sheep, and others, were also inventoried. Spoons, watches, mugs, china, tables, chairs, beds, tools, and everything in between were listed and accounted for. In total, his possessions at Ferry Farm were worth 824 pounds, 8 shillings, and 3 pence.[36] To compare, five years earlier, he had bought nearly 300 acres of land for a quarter of that amount. That land was not cheap.

AS RICH AS AUGUSTINE MAY HAVE BEEN, HIS WEALTH WAS NEARLY INSIGNIF-icant to such families as the Fairfaxes. In 1719, Lord Baron Thomas Fairfax, about twenty-six years old, owned roughly five million acres—approximately 90 miles squared, or 7,800 square miles—of land in the Northern Neck.

The Washington family was placed comfortably in the "the second tier of gentry, below the colony's royal officials and its governor," while placing themselves in, according to Mount Vernon's researcher Mary Thompson, "the wealthiest tenth of Virginia's population."[37] The Fairfaxes may have seen the Washingtons as simple gentry, but to the other 90 percent of Virginia, they were a name unto themselves.

GEORGE WASHINGTON PARKE CUSTIS NOTED THAT ON HIS DEATHBED AUGUS-tine was said to have uttered, "I thank God that in all my life I never struck a man in anger, for if I had I am sure that, from my remarkable muscular powers, I should have killed my antagonist, and then his blood at this awful moment would have lain heavily on my soul. As it is, I die in peace with all mankind." How much of this is paraphrasing from Custis is unknown, but his humble-bragging would not have been unusual for a man of such stature.[38]

Mary Washington was about thirty-seven years old when she lost her husband. Marriage was a sacred institution in the colonies. Annulment was possible, in which case, legally, the marriage never took place. But that was a rare circumstance. Marriage was serious. "When [a woman] said, 'Until death do us part,' she meant it. Divorce was unknown; its possibility undreamed of."[39]

Unlike other women, especially of childbearing age, Mary never remarried. "Mary Washington was loyal to her husband's memory and to his trust. And now, having to assume her husband's duties in addition to her own, no time for sorrowful brooding was permitted to the widowed mother."[40] Perhaps she didn't remarry so that she could keep her inheritance willed from Augustine; if she remarried, all land, including the children's, would be the new head of the family's to manage as he saw fit. All that Augustine inherited and worked for, and all that Mary inherited, would have gone to this new man. This is the view of Michelle Hamilton, manager of the Mary Washington House in Fredericksburg. She emphasized that this was not a selfish move on Mary's part just to keep the land or keep power over a plantation: "She put her children ahead of herself." It was not a detriment to Mary's character—the opposite of her relatively late age when first marrying—to remain a young widow. "Mary Washington was a strong woman who managed a plantation and five children on her own in a world dominated by men. . . . This accomplishment should be respected."[41]

The custom of remarrying was so common, for both men and women, that it can't be emphasized enough how unusual it was that Mary never remarried. In colonial Maryland, in Charles County, two-thirds of widows and widowers remarried within the first year after their spouse's death.[42] It was a necessary custom for those who could not raise either a plantation or children alone, just as Augustine had done.

Mary could. Despite possible whispers and rumors from nearby neighbors about her widowhood, whether about her personality or about possible courtiers grabbing a piece of the Washington land, she did not care. In a way, it put her above many other women of her time. She lived in a man's world, so a young widow with children and a plantation would naturally have an uphill battle to fight.

When Augustine died, so did whatever dependence Mary had. Though she was plagued in her early girlhood by the deaths of her father, mother, guardian, sisters, and others, and though she bounced around from place to place, she was still relatively looked after. Her family members' wills made sure of that, and her siblings were almost a generational difference in age. When Augustine died, she did not have another to look up to, to take the brunt of responsibility. Now she had to manage a plantation, children, and slaves, all at the age of thirty-seven.

Augustine's death was her baptism by fire.

Chapter 7

Matriarch

THE WIDOWHOOD AND MIDLIFE OF MARY WASHINGTON

1743–1754

---

*"Saddled with responsibility for the four younger siblings,
the boy was hostage to her whims and steely will. . . ."*

The patriarch was dead; long live the matriarch. Johann Wolfgang von Goethe, a poet and contemporary of George Washington on the other side of the Atlantic in Frankfurt (then an imperial free city within the Holy Roman Empire), wrote in his work *Wilhelm Meister's Years of Travel,* "She is the most excellent woman, who when the husband dies, becomes as a father to the children."[1] That is exactly what Mary Washington was to her children: a mother and a father, fulfilling both maternal and paternal roles to both her step- and biological children, adult and young.

The Washington family's entire structure was upended by the death of Augustine. It fundamentally changed the future of the children—George included. One historian, Richard Norton Smith, said that with the loss of Augustine and under the guardianship of his mother, "George Washington grew into an emotionally inaccessible man who channeled his considerable passions into, first, self-advancement, and second, building a nation."[2]

Many years later, George Washington Parke Custis, the grandson of Martha Custis, née Dandridge (later Martha Washington), and the step-grandson of George Washington, wrote that Mary, "by the death of her husband, became involved in the cares of a young family, at a period when these responsibilities seem more especially to claim the aid and control of the stronger sex; and it was left to this remarkable woman, by a method the most rare, by an education and discipline the most peculiar and imposing, to form in the youth-time of her son those great and essential qualities which led on to the glories of his after-life."[3]

One author, Paul Ford, writing in 1896, went in the opposite direction from George's descendant: George's mother had nothing to persuade or offer him in these years. "The sentimentality that has been lavished about the relations between the two and her influence upon him, partakes of fiction rather than of truth. . . . The boy passed most of his time at the homes of his two elder brothers, and this was fortunate."[4] Factually, again, this was more or less accurate, but it's wrong to think that Mary was nothing to her son during these years. As George himself wrote, it was Mary by whose "Maternal hand (early deprived of a Father) I was led to Manhood."[5] Indeed, it would be outlandish to suggest that *General* George Washington did not inherit some of his confidence and poise from his "commanding" mother. As his cousin Lawrence attested, "Whoever has seen that awe-inspiring air and manner so characteristic in the Father of his Country, will remember the matron as she appeared when the presiding genius of her well-ordered household, commanding and being obeyed."[6]

These were the years in which every decision made by Mary could alter his path, or his career, away from the future generalship of the Continental Army and the future presidency of the United States of America. Whether Mary intended it or not, these years were to shape the future for George.

By extension, they were to shape the future of the colonies.

AUGUSTINE WASHINGTON WAS BURIED AT THE BRIDGES CREEK PROPERTY, later called Wakefield, shortly after his death, in the family crypt. Today a crypt stands together with the vaults of his parents and grandparents, bearing the name of the father of George. Over thirty of the Washington family members' grave markers still stand there.

Augustine's will was probated in early May of 1743 by Lawrence, his eldest, placing the will in legal effect and thus lawfully allowing Mary to have the control of his property. Their children who had not come of age were now under Mary's jurisdiction. She was left with full custody of the children, as well as the property that they inherited. This was not an unusual move in the eighteenth century. The most challenging issue was how to afford the children, with education being a priority in many husbands' wills. "Raising children alone could be an expensive and difficult responsibility for a widow," wrote historian Daniel Smith.[7]

THE IMMEDIATE EFFECT OF THE DEATH OF AUGUSTINE WAS APPARENT, AS Mary was forced to pull George out of his formal education. Both Lawrence and Augustine Jr., George's older half siblings, received formal education in England, at the same Appleby School that their father attended. Child George met them both when the two brothers returned sometime in 1738, when the Washingtons were isolated in Little Hunting Creek. Esteemed historian Douglas Freeman noted that George "quickly made a hero of Lawrence," who had the "bearing and manners that captivated" the young impressionable boy.[8] Indeed, it later became evident how that hero worship and emulation applied to the teenage George as he matured.

Mary could not afford to send George to England, whether or not she wanted to, considering the conflicts that plagued Europe. During the 1740s, England was in the midst of a war with the French, Spanish, Bavarian, Swedish, Russian, and other monarchies and dukedoms. It would be called the War of the Austrian Succession and would end in 1748. It was a complicated series of battles and conflicts over whether Maria Theresa, daughter of the deceased emperor Charles VI, had the right to lay claim over Austria. As continental as the conflict seemed, England was hit hard in 1745 by an insurrection of Jacobites, who demanded the restoration of the Stuart dynasty, whose army invaded England to the panic of many. Ultimately, the Jacobites were defeated in 1746. (Among the possible rebels was Hugh Mercer, friend of Washington, physician, apothecary, and surgeon, who fled to the colonies but settled in Fredericksburg, Virginia, decades later.)

Mary may have been aware of all that was happening and decided England was no place for her young, eleven-year-old boy. Surely if it wasn't appropriate for him, it wasn't for her other sons, either. Samuel was nine years old by the time of his father's death; John Augustine, seven; and Charles, five. John Augustine and Charles were still too young to think about formal education, but the situation in Europe only deteriorated in the next decade.

Mary perhaps saw no need for formal education overseas, with George's older half siblings already having completed their schooling. Her own education was uninspiring, being only average—or worse—for a woman in the early eighteenth century in Virginia. In the years following, wrote Nancy Turner, "though more attention was being paid, by this time, to the matter of education, opportunities and resources were still meagre in the colonies. People continued to regard knowledge as a gift rather than an attainment."[9]

Scant facts exist about the education of George Washington. Even fewer exist for his younger siblings. But there was some education, and certainly it was not equal to the expense or the extent of that of his older brothers. The instructor was a sexton of a local parish, named Hobby, rumored to be a former convict, who had arrived from England with Augustine. For the next two years, George was taught by a private tutor, Williams, farther north by Popes Creek. Upon returning to Fredericksburg in 1738, he and his four younger siblings were instructed by Reverend James Marye. A Frenchman, Marye in 1726 renounced Catholicism and his Jesuit priesthood education, fleeing to England, where he married Letitia Staige, sister of the rector at Saint George's Parish in Fredericksburg. Soon after 1728, Marye and his wife arrived in Virginia, where he was a minister for French Protestants. He was deemed popular enough to be assigned to Saint George's Parish in October 1735.[10]

Bishop William Meade, a century later, said that "Mr. Marye was a worthy exception to a class of clergy that obtained in Virginia in olden time. So far as we can learn, he was a man of evangelical views and sincere piety. We have seen a manuscript sermon of his on the religious training of children, which would do honour to the head and heart of any clergyman, and whose evangelical tone and spirit might well commend it to every pious parent and every enlightened Christian."[11] Under Marye's tutelage, George was taught the basic education and morals needed for the time. The only authorized biographer during Washington's life, David Humphreys, wrote that "he was betimes instructed in the principles of grammar, the theory of reasoning, on speaking, the science of numbers, the elements of geometry, & the highest branches of mathematics, the art of mensuration," among other studies.[12]

When George was thirteen, only a short two years after his father's death, he learned surveying, a necessary skill in the Virginian planter's life. He was also learning geography and geometry, without which surveying landscape would be impossible.

"SURVEYING," he wrote in August 1745, in what was suggested to be his earliest writing, in clear, elegant script, "Is the Art of Measuring Land and it consists of 3 parts. 1st. The going round and Measuring a Piece of Wood Land. 2d Plotting the Same and 3d To find the Content thereof and first how to Measure a Piece of Land."

This same schoolbook, in his geography lesson, contained a list of the major colonies, provinces, and islands of North America. This includes "Colofornia," "Porto Rico," and "the Caribbee Iselands."[13]

THE EDUCATION THE YOUNGER WASHINGTON CHILDREN RECEIVED WAS INferior in both quality and quantity to that of Augustine's first children. Take, for example, the knowledge of Latin or Greek. There were no in-depth lessons for George or the others. What use would the language of the ancient Romans or Greeks be to plot a piece of land? Would Cicero have helped him when measuring the frontier? Certainly not. The skills he learned were egalitarian. This was used against George decades later when Reverend Jonathan Boucher, the fierce Loyalist and former tutor of George's stepson John Parke Custis, wrote with a hint of elitism that George "like most people thereabouts at that time, had no other education than reading, writing and accounts."[14] How infuriating to the Loyalists and royalists that a simple farmer, in their eyes, with no college education, was leading this revolution. George's education stopped when he was fifteen, a mere teenager, typical for the time.

How, too, George and his younger siblings would have changed had his father not died or had his mother, Mary, decided not to keep them in Virginia. Had he been educated in England, he would have come to sympathize with the European way of thinking. Perhaps seeing London and the House of Parliament and St. James's Palace would have struck him not as excessive and authoritative, but as a rich and proud display of power for the mighty British Empire, especially against the Spanish or French.

This was the first of many of Mary's decisions to shape him. Per Jared Sparks, "There never was a great man, the elements of whose greatness might not be traced to the original characteristics or early influence of his mother."[15]

THE INTERVENING YEARS, FROM 1743 AT THE DEATH OF AUGUSTINE, TO THE year 1758 when George Washington resigned from the Virginia militia in the French and Indian War, were ultimately what shaped the man to become the rebellious tactician and general who the war-torn colonies sorely needed decades later, in the opinions of many, including historians Douglas Freeman and Ron Chernow.

This decade shaped him more than any other in the antebellum colonies. Not just because of his war experience in Ohio or the failed Braddock Expedition, or his experience in surveying, or the budding relationship with a woman named Martha Dandridge, but also through his relationship with his mother.

"Saddled with responsibility for the four younger siblings, the boy was hostage to her whims and steely will. . . ."[16]

IT WOULD BE A BLUNDER TO TALK ABOUT GEORGE'S UPBRINGING WITHOUT mentioning his meticulous copying of 110 maxims of civility. The work, currently in the Library of Congress, is titled *Rules of Civility & Decent Behavior in Company and Conversation*. It was a collection of one-sentence advice for formal and social events and good manners that young George clearly took to heart. George had written the rules in his schoolbook when he was a young teenager; the ten-page exercise, even when taken alone, provides a peek into his character.

1. Every action in company ought to be with some sign of respect to those present. . . .

8. Let your discourse with men of business be short and comprehensive. . . .

11. Shift not yourself in the sight of others nor gnaw your nails. . . .

15. Mock not, nor jest at any thing of importance; break no jests that are sharp-biting, and if you deliver any thing that is witty and pleasant, abstain from laughing thereat yourself. . . .

23. Utter not base and frivolous things among grave and learned men; nor very difficult questions or subjects among the ignorant; nor things hard to be believed. . . .

24. Be not immodest in urging your friend to discover a secret. . . .

39. Let your recreations be manful, not sinful.

And so on, for over one hundred items.[17] These rules were adopted from French Jesuits at the end of the sixteenth century, who were connected to one of young George's own teachers, James Marye.

"What, then, made Washington the greatest man of this great generation?" asked an article in the *Daily Signal* in 2015. "It was his character." He was autonomous, able to make quick decisions on rationality and not on "undisciplined passion, whether a personal passion, like pride, or an intellectual passion."[18] And though this same article says he self-taught these virtues, it was his mother and older half brother who planted the seeds.

As influential as his mother's devotional texts were when he was young, especially Sir Matthew Hale's *Contemplations, Moral and Divine* (he kept it in his library with his own bookplate), these short teachings stuck with him. Perhaps Mary imparted some of these wisdoms and manners to the little boy, to maintain the status of the Washington family. His own maxims were published in book form mere decades after his death, taking phrases from letters or speeches, and compiled according to topic by scholars; his book was a clear mirror of these rules.[19]

A TALE AS TOLD BY GEORGE WASHINGTON PARKE CUSTIS SHOWED A DIFFERent side of Mary, not one of authority but of comforter. At an unknown age, but surely after Augustine's death, teenage George and a few friends tried to saddle and place a bit in the mouth of a prized horse, beloved by Mary. It was a "blooded horse" that Augustine admired, which "was the Virginian favorite of those days as well as these." This horse had never before been ridden, and young George, cocky as he was, believed it to be time. The horse did not, and through an ensuing fight of bucks and kicks, apparently burst a blood vessel and immediately died.

Soon after, at morning breakfast, Mary asked the children, "Pray, young gentlemen, have you seen my blooded colts in your rambles? I hope they are well taken care of; my favorite, I am told, is as large as his sire."

George replied, without missing a beat, "Your favorite, the sorrel, is dead, madam."

"Dead. Why, how has this happened?"

"The sorrel horse has long been considered ungovernable, and beyond the power of man to back or ride him; this morning, aided by my friends, we forced a bit into his mouth; I backed him, I rode him, and in a desperate struggle for the mastery, he fell under me and died upon the spot."

Mary was flustered. But as fast as anger or sadness arose in her, it disappeared. She said, slowly, "It is well; but while I regret the loss of my favorite, *I rejoice in my son, who always speaks the truth.*" The italicized was in Custis's original text.[20]

This story was the equivalent—a part two, perhaps—of the morals of George's honesty. Instead of his father and the cherry tree, it was the mother and her horse. The story is nearly identical: George, full of wonder, brazenly committed an act that destroyed some prized possession; the parent asked what happened; George admitted his culpability. But here, without a father, the mother took this role. Much of the responsibility fell on her to teach, comfort, praise, and in times of need reprimand.

THERE WERE OTHERS WHO INFLUENCED GEORGE AS WELL, WHILE HE WAS under Mary's care. From the first time George met his older, wiser half brother Lawrence, the heir to the Washington estates, the younger boy admired him more than anyone else in his early life. After the death of Augustine, Lawrence himself became that missing father figure so needed in the child's and teenager's life.

Lawrence certainly had the reputation of a Washington, perhaps even exceeding the reputation of his father or grandfather. He was a veteran of both the War of Jenkins' Ear from 1739 to 1748 and King George's War from 1744 to 1748. While both conflicts were related to the War of Austrian Succession that plagued Europe, it was King George's War that was part of the French and Indian War, a decades-long series of on-and-off conflicts and battles over territory in the Americas. In the former, the War of Jenkins' Ear (which only received its name a century after the fact), Lawrence was promoted to captain, leading a Virginia company under Admiral Edward Vernon in March 1741 at the ill-fated Battle of Cartagena (modern Colombia), in which Spanish forces overtook the region south of North America and staked their claim to the continent. The battle was not unknown, appearing in the *Virginia Gazette*.[21] Of the nearly four thousand colonists who volunteered, only six hundred returned, the others mostly dying from disease. Lawrence was one of the lucky few, and in honor of Admiral Vernon's gallantry, renamed Little Hunting Creek as "Mount Vernon."[22]

In July 1743, Lawrence married fifteen-year-old Anne Fairfax, ten years his junior and the daughter of the rich and prominent Colonel William Fairfax, who owned five million Virginian acres and lived in Belvoir, just south of the Mount Vernon estate. The marriage was not without scandal, however, as Major Washington—he was promoted a few months earlier—publicly accused Reverend Charles Green of Truro Parish of trying to defile (in the words of Lawrence, "debauch") the young unmarried woman. This same Charles Green had been nominated as rector by Augustine Washington a decade earlier. Green denied it and countersued Lawrence for slander. Eventually this became the "only full-scale ecclesiastical trial" in the history of the Colony of Virginia. Evidence was presented, including from Anne's stepmother, Deborah, who claimed that years earlier Green fondled Anne, who was only nine years old then. She also claimed the reverend, five years later, attempted to rape Anne. Reverend Green's sister-in-law painted a different picture, one of Anne being a young seductress who willingly and willfully flirted with her brother. The scandal was so enormous that even the governor of Virginia, William Gooch, had to step in, making a deal between these two families: Green would not be defrocked or removed as Truro's rector, but he would drop the slander accusation and pay for court costs. "But on the key point," says Peter Henriques, who discovered these documents in the Library of Congress in 1990, "Green won."

The matter was finally settled in the spring of 1746 and everyone eventually moved on. George and Green, through the following decades, wrote numerous correspondences to each other.[23]

The nine-year marriage between Lawrence and Anne was a relative failure, as far as society was concerned: they conceived no children who would survive past childhood before his early death.

And still, through all of this, George looked up to Lawrence. Scandals and rocky marriages or not, Lawrence was a mentor, brother, and much needed older figure for George. Perhaps this was a port in the storm of Mary. George "had plenty of young companions," wrote Charles Moore. This included his other siblings, and members of his aunts' and uncles' families and their children. They were spread out in age, though all lived relatively close to George.[24] Betty Washington was a year younger than George; Samuel, two years younger; John Augustine, four years younger; Charles, six. All were companions for young George as he grew up. This was in sharp contrast to his mother's childhood, where siblings were too removed in age to be considered siblings at all. But among his family's servants, his siblings, his cousins, his aunts and uncles, and his neighbors, Lawrence was his favorite.

Which is precisely why he wanted to follow Lawrence into the British Royal Navy.

ONLY A FEW SHORT MONTHS AFTER THE TRIAL, IN EARLY SEPTEMBER OF 1746, that the young George found his true calling—or, at the very least, the fourteen-year-old believed that he did. He wanted to be like his brother and join the British Royal Navy, go on seafaring adventures close and far, become a seaman. What he, Lawrence, or Mary said to him is lost to history, though clues exist from three specific letters from other figures.

On September 9, 1746, William Fairfax wrote to Lawrence, "The weather being so sultry, and being necessarily obliged to go about this town to collect several things wanted, I have not yet seen Mrs. Washington. George has been with us, and says He will be steady and thankfully follow your Advice as his best Friend. I gave him his Mother's letter to deliver with Caution not to shew his. I have spoke to Dr. Spencer who I find is often at the Widow's and has some influence, to persuade her to think better of your advice in putting Him to Sea with good Recommendation."[25]

Presumably George's one letter to Mary would have been more diplomatic, perhaps even yielding to Mary's ultimate decision. It was almost a tale of brothers keeping something hidden from their mother, lest she ruin the fun or disapprove. This particular instance revealed a great deal about George's early relationship with his widowed mother.

Within the following week, the news was told to Mary of George's plans. George was eager enough to have his bags packed to travel, or even to be onboard the ship itself.[26] She may have agreed . . . until she didn't. Robert Jackson, a Washington family friend and Fredericksburg neighbor, said to Lawrence on September 18, just a week after the first letter from William, that he was "afraid Mrs. Washington will not keep up to her first resolution," that first resolution being approval. Then she thought it over. Well, he said, Mary "seems to intimate a dislike to George's going to Sea and says several persons have told her it's a very bad Scheme."

Jackson continued, "She offers several trifling objections such as fond and unthinking mothers naturally suggest, and I find that one word against his going has more weight than ten for it." "Unthinking" to him, perhaps, but not to Mary. She would give it much thought. He promised to talk to her in the very near future, knowing full well that this was a high priority for everyone involved.[27]

Again, historians are left to wonder what happened in the conversation.

The controversy did not die until the spring of 1747, about eight months later, when Mary received a reply to her letter from her brother Joseph in England. In these intervening months there may have been conversation among George, Lawrence, and Mary.

In the May letter from Joseph, he confirmed all her worst fears: "I think he had better be put apprentice to a tinker, for a common sailor before the mast has by no means the common liberty of the subject; for they will press him from ship to ship, where he has fifty shillings a month, and make him take twenty-three, and cut and slash and use him like a negro, or rather like a dog." The navy was abusive to young boys like him. The conditions were horrible. But that wasn't all. Joseph continued, "If he should get to be master of a Virginia ship (which is very difficult to do), a planter who has three or four hundred acres of land, and three or four slaves, if he be industrious, may live more comfortably, and have his family in better bread than such a master of a ship can."[28]

Strike two and strike three. Abusive, dead-end, and ultimately pointless for a Washington.

How did George react to his mother's decision? He would have surely thought of rebelling against his mother—he was a rebel in later life, after all; why not during his teenage years? Any conversation between mother and son could have seesawed between begging and anger and hurt, involving a mother whose son was abandoning her for a life of danger and abuse and a son whose mother wanted to change and control his calling. The relationship they had would have been tarnished, maybe even irrevocably.

JOSEPH'S ACCUSATION OF ABUSE WAS NOT A WILD EXAGGERATION. ONLY forty-two years after this letter, the famed mutiny of the crew on the HMS *Bounty* occurred when they seized the ship from Lieutenant William Bligh, partly because of his abuse against the crew in the name of "discipline."

If George joined the Royal Navy, and any infraction was held against him at sea, then the Washington name would be associated with his actions. Punishment was very public, either whipping, flogging with a cat-o'-nine-tails, or other means, such as cutting back on rations of rum. It was the ultimate deterrent.

Further, rumors of sexual abuse from the older and senior seamen were rampant. Indeed, instances of punishment exist during this time for not just homosexuality and "buggery," but also pederasty. A twelve-year-old boy, John Booth, accused one Henry Brick in 1757 of raping him in bed. It was potentially a capital offense; after trial, Brick was instead sentenced to five hundred lashes and forcibly discharged from the navy.[29] Other instances and trials exist of pedophiliac sexual abuse (for boys even younger than George at the time), so this was not a onetime offense, but relatively common, and certainly a concern.

Mary, when inquiring, very well could have heard about these predators at sea. The horrid image of a son being abused would have sent any mother into hysterics.

The letter from Mary's brother Joseph solidified Mary's resolve. Mary refused to honor George's request.

Though George had allies among both friends and family, he accepted his mother's wish, if reluctantly. She had legitimate fears, not just for his safety but for the well-being of her family. "This was her eldest son," wrote Jared Sparks, in defense of George's mother, "whose character and manners must already have exhibited a promise, full of solace and hope to a widowed mother, on whom alone devolved the charge of four younger children."[30] Mary's disapproval was blistering. "Of the mother," said George's cousin Lawrence, who frequently visited his family at the Fredericksburg farm, "I was ten times more afraid than I ever was of my own parents. . . . I have often been present with her sons, proper tall fellows too, and we were all as mute as mice." Yet she was still George's mother, and could still show affection. "She awed me in the midst of her kindness," Cousin Lawrence continued, "for she was indeed truly kind." She was "commanding," he noted, an apt description for the mother of the future commander of the Continental Army, and he "could not behold that remarkable woman without feelings it is impossible to describe."[31]

To some historians, this was the fracture of the Mary-George relationship. This was when she turned into a bitter authoritarian. "She would always be strangely indifferent to his ambitions," wrote Washington biographer Ron Chernow. This, he said, was a selfish act, putting her own wants above that of her son's.[32] Custis saw it a different way: His great-grandmother instilled obedience into George. Less authoritarian, more maternal authority—an important distinction. She "taught him the duties of obedience, the better to prepare him for those of command. . . . The matron held in reserve an authority, which never departed from her."[33]

HOW WOULD THIS HAVE CHANGED THE BOY? LIKE HIS HYPOTHETICAL EDU-cation in England, it would have broadened his horizons. Going to sea at the command of a Royal Naval ship would have brought a distinctly different perspective than the Virginian planter's lifestyle that he knew. Would history have had Captain George Washington fighting alongside Admiral of the Fleet Sir George Martin, blockading rebel ports in the continent? Perhaps, but he was called to be more than a captain of a ship or a midshipman.

Benson Lossing, with a religious twist that Mary surely would have appreciated, put it eloquently: "He was destined by Heaven for a far nobler career than man had conceived for him. This incident illustrates the truth of the familiar apothegm, 'Man proposes but God disposes.'"[34]

Mary's fear of losing her oldest child was well-founded. She had already lost Mildred, her second daughter and youngest child. She had lost her husband. She lost both parents and multiple guardians. Some of her stepchildren had died prematurely, far before their time. However, it was several years later when George took his one and only trip outside the continent that perhaps proved to Mary that her decision was not only a mother's intuition, but real-world wisdom.

In late September 1751, nineteen-year-old George and his companion-brother Lawrence sailed off to Barbados. Lawrence had tuberculosis and hoped to find a cure in the tropics away from the upcoming cold Virginia winter, following months of failed remedies. He took his younger brother with him, in his only venture outside of the American continent. It was perhaps a concession to the young lad, who surely must have been sad, still, not to have a life at sea.

The voyage was rough, with high winds and a rocking ship. For five weeks they were "toss'd by a fickle and Merciless ocean," and any report of a steady day after weeks of seasickness was welcomed news. By November 2, they landed, sooner than George had expected.

Not long after, George contracted smallpox, with all the classic signs of a fever and infection. It was the great destroyer and plague of the Native Americans, whose immune systems had never encountered such a disease. Years of exposure left many tribes in North, Central, and South America dead. It was their Black Death, of even greater proportions. For Europeans, it was still fatal, though not nearly at apocalyptic scales. It "reportedly disfigured, crippled, or killed every tenth person" during the eighteenth century; an outbreak occurred in Williamsburg only three years earlier, infecting nearly eight hundred people, and killing, of those, over fifty.[35]

And George contracted it.

"Was strongly attacked with the small Pox," he wrote on Saturday, November 17, in his diary.[36] For the next month, until December 12, his diary remained blank, presumably as he recovered. He had contracted the disease sometime two weeks earlier, possibly on November 4, when he had dinner at Gedney Clarke's residence. "We went," George wrote that day, "myself with some reluctance, as the smallpox was in his family." In an ominous note that same day, George wrote that Gedney's wife, Mary, "was much indisposed, insomuch that we had not the pleasure of her company."[37] Perhaps it did not matter whether he came in direct contact with her; the Clarke family passed it on, symptoms first showing in George thirteen days later.

The purpose of the trip—to heal Lawrence—failed, and George's own disease left him seasick and scarred. He never left the country again. It did give him a lifelong immunity, however, a beneficial advantage decades later during the Revolutionary War, when smallpox was rampant among his troops (it was so bad, in fact, that in 1777 it was required that all recruits be inoculated before joining the Continental Army).[38] For the foreseeable future, though, it validated Mary's fears of her boy's adventures at sea.

It surely vindicated her; and to George, that was his end of that fleeting dream.

WITH MARY'S REFUSAL TO GEORGE'S SEAFARING REQUEST IN 1747, THE BOY looked elsewhere to fulfill a life of adventure that he so wildly needed. He often saw his brother Lawrence at Mount Vernon, the former Little Hunting Creek plantation, which provided an escape from his mother. This was his first step beyond home. John Locke had noted as much, that a child "will, perhaps, be more innocent at home, but more ignorant of the world, and more sheepish when he comes abroad."[39] This was him outside his shell.

George decided to take up surveying, using his newfound connection to the local Fairfax family to jump-start his way. Surveying was a profitable, high-demand career. "Educated young men on their way up in the world often took to surveying as an entry-level profession. They started as apprentices before moving on to freelance work or salaried appointments."[40] George's skills excelled, and within years he was an expert in the field. In July of 1749, he surveyed Alexandria, Virginia, far north on the bank of the Potomac River.[41]

Truly, Lord Thomas Fairfax saw great potential in George's prospects, as within a few short months, right after his sixteenth birthday, in March 1748, he and his friend George Fairfax, Lord Fairfax's cousin's son, were off to the western Virginian mountains on a surveying party. Here he had on-site experience with a circumferentor, or compass, and other instruments, learning triangulation and geometry. This was the first breath of expansion that the teenager had, outside of the family moving to and from houses when he was little. George made meticulous diary entries every day—some more exciting than others. On those less exciting days, he nonchalantly wrote "nothing remarkable happen'd." On Tuesday, March 15, he wrote that the party "Worked hard till Night," before going to bed on "nothing but a Little Straw." The next day, he enjoyed a nice feather bed in modern-day Winchester, Virginia.[42]

Back at Ferry Farm, Mary perhaps had second thoughts about the path she forced upon her oldest. Lawrence, then directly cultivating the boy, probably wanted to send him to England for education. Maybe, he thought, international experience was a great idea, no matter what Mary said. Certainly the interest of family friend Thomas, Lord Fairfax, in George would offset any financial worries. Mary wrote a letter to Fairfax at Belvoir, asking his advice as to what the young man, by that time living away from his mother, could or couldn't do. He responded some time later with an examination of the young man, his abilities and personality:

You are so good as to ask what I think of a temporary residence for your son George in England. It is a country for which I myself have no inclination, and the gentlemen you mention are certainly renowned gamblers and rakes, which I should be sorry your son were exposed to, even if his means easily admitted of a residence in England. He is strong and hardy, and as good a master of a horse as any could desire. His education might have been bettered, but what he has is accurate and inclines him to much life out of doors. He is very grave for one of his age, and reserved in his intercourse; not a great talker at any time. His mind appears to me to act slowly, but, on the whole, to reach just conclusions, and he has an ardent wish to see the right of questions—what my friend Mr. Addison was pleased to call 'the intellectual conscience.' Method and exactness seem to be natural to George. He is, I suspect, beginning to feel the sap rising, being in the spring of life, and is getting ready to be the prey of your sex, wherefore may the Lord help him, and deliver him from the nets those spiders, called women, will cast for his ruin. I presume him to be truthful because he is exact. I wish I could say he governs his temper. He is subject to attacks of anger on provocation, and sometimes without just cause; but as he is a reasonable person, time will cure him of this vice of nature, and in fact he is, in my judgement, a man who will go to school all his life and profit thereby.

I hope, madam, that you will find pleasure in what I have written, and will rest assured that I shall continue to interest myself in his fortunes.[43]

Fairfax indeed kept his promise and took a keen interest in the boy, more so than any of the Washingtons, perhaps even more than in Lawrence the heir.

WHATEVER DIFFICULTY IN THE RELATIONSHIP BETWEEN MOTHER AND SON, George still continued to care for his mother. On July 15, 1748, while visiting relatives in Gloucester County, George made a stop in nearby Yorktown—about ninety miles southeast of Fredericksburg—and shopped for his mother. There he bought ribbons from a Mr. Mitchell for 1 pound, 3 shillings, and a glass ring for 3 shillings and 4 pence. The money had been provided specifically from his mother.[44] Only a week before, 165 acres of Ferry Farm were sold for 110 pounds. George was aware of this deal and presumably approved—or, at least, he did not object.[45] Nevertheless, Mary continued to provide George financial support for both his and her needs when applicable.[46]

THE RESPONSIBILITIES OF WIDOWHOOD AND OVERSEEING PROPERTY MUST have hit her when she received the estate's bills. Financial matters could no longer be swept away for someone else to handle; they were her duties now.

For instance, in June of 1751 and January and April of 1752, Dr. Sutherland treated various members of the Washington estate. On April 3, he helped a "negro fellow" drain a tooth, for which he charged Mary an exact 2 shillings, 5 pence. Other slaves were treated, and even John Augustine and George on one occasion, for various ailments. The total bill for these visits came to 1 pound, 1 shilling, and 5 pence.[47]

Another account a decade later showed she paid James Buchannen over 25 pounds for a month's worth of material. An extremely deteriorated receipt shows Mary had purchased in July

1765 "110 yds of [destroyed]," "3 Gallons of [destroyed]," and so on. The total sum of these purchases certainly added up.[48]

Mary, as a plantation owner, had to take care of her property, always at her expense. She had to provide for her slaves and her family, and herself. She had to keep herself in the gentry status. For a woman whose infamy in history paints her with an almost obsessive interest in money, this was a drain on her ambitions, whatever they may have been.

AS IF THE CONSTANT EXPENSES OF THE PLANTATION WERE NOT ENOUGH, tragedy struck the Washington family in July of 1752, when Lawrence Washington, heir apparent, died at about the age of thirty-three. He never recovered from the tuberculosis that haunted him for years, and remedy after remedy failed him. It was less than a year after he went to Barbados, and his return trip just a month before his death was clearly an acceptance of his fate. He was going to die. He knew it, and his family knew it. He and his wife, Anne Fairfax, had no children who lived past four years old. Most died within, or soon after, their first year. As heroic as Lawrence seemed to be to George, he was still mortal.

When his half brother, his second father, died, George received condolences from friends and family. "I most heartily condole You on the Loss of so worthy a Brother & Friend," wrote a neighbor.[49] George and Lawrence's relationship had never dwindled, and the brothers had become confidants. George complained of Mary's refusing to build a house on Deep Run tract for herself, in accordance with her husband's will, and instead staying at nearby Ferry Farm.[50] The land was only hers until George turned twenty-one, which was fast approaching. George trusted Lawrence, in part because he was the oldest of the generation and had more sway with Mary (when she wasn't obstinate).

In Lawrence's will, he designated his wife to be heir of the Mount Vernon estate; upon her death, it should pass to his daughter, Sarah.[51] Sarah, however, did not live past four years old, dying only two years after her father. Since Anne had already remarried to Colonel George Lee and had no use for the Mount Vernon land, she agreed to rent it out to George that same year.

From here on out, George and Mount Vernon would be permanently connected, only broken in times of war or during his presidency. George moved out of Ferry Farm to Mount Vernon—where he had often visited—in 1754, marking a new beginning of his life.

Mary was now separated from her oldest son, the same son whom she wanted to protect—or order around—so badly. In many ways, it was a new beginning for her life as well.

Chapter 8

Lieutenant Colonel Washington

CLASH OF THE TITANS: ENGLAND AND FRANCE

1754–1772

*"I heard Bulletts whistle and believe me there
was something charming in the sound."*

The year 1754 was an unusual year not just for the Washingtons, but for the entirety of North America. For the
past seventy years, France and England had laid claim to
different, or sometimes the same, land on North American soil.
By 1754, this was resolved through three overarching phases: in
the late 1600s was King William's War, in the early 1700s were
Queen Anne's War and Dummer's War, and in the 1740s were
King George's War and the War of Jenkins' Ear. These conflicts
were by-products of the overall international problems.

The year 1754 would be the start of the final phase. Though
history knows the collective conflicts as the French and Indian
Wars, this final one, lasting nine years, would be known to the
English as *the* French and Indian War. As a theater of the Seven
Years' War in Europe—very much a world war—from 1756 to
1763, this colonial conflict was to shape future imperial-colonial
relations.

The Ohio Valley was again up for grabs between the English
and French Empires. It was prime real estate for both, as settlers
wanted to expand westward.[1] In the meantime, George had joined
the Virginia militia, becoming a major. His boyhood education

paid off, as he had been personally selected by Governor Robert Dinwiddie as emissary to the French. He was tasked in late 1753 to deliver an ultimatum to the French: leave the Ohio or be attacked. The conversation he had with his mother would have been a repeat of before: "Do not go, my son, or you will die." Yet he was prepared, and was reported to have said in reply, "Madam, didn't you above all others teach me the importance of duty?"[2]

During the 1600s, Native Americans had called George's great-grandfather John Washington "Conotocaurious," or "Town-Destroyer." The French-aligned Iroquois gave the same name to then Lieutenant Colonel George Washington. After leading a series of battles, Washington was captured at Fort Necessity in July of 1754, and subsequently released, returning to Virginia. A month prior to the news of his capture, he was hailed as a hero, and news reached his mother and the Fredericksburg area. "Your Mother &c. whom I frequently see are well, very lately I had the honour to dance with her, when your health was not forgot," wrote Daniel Campbell, a Fredericksburg friend and fellow Masonic member.[3] The same day, William Fairfax also wrote to George, saying that "Yr Mother & Family are well and Send their Several Greetings, desiring often to know of yr Welfare & Progress."[4] His welfare and progress were grand, truthfully, but only for the time being; within a month, he was captured. Heroics did not mean infallibility, and George's failure at Fort Necessity persuaded him to resign his commission in October of that year.

In these months, George grew beyond, perhaps, what his mother would have wanted. He killed men and led men to kill and to die. Indeed, his actions and the death of one man directly led to France and England formally declaring war.

Joseph Coulon de Jumonville was a French-Canadian officer who died in late May 1754 at the Battle of Jumonville Glen in current-day Pennsylvania. He was killed by George's forces. The French claimed that he was to deliver a peaceful message to the British—no more than a messenger. But the British believed he was leading his thirty-five men to spy on nearby Fort Necessity. Washington deployed his men, and the two forces fought. Joseph Coulon was wounded and captured. However, the chief of the British-allied Iroquois who accompanied Washington in battle, named Half King, killed Joseph with a tomahawk strike to the head.

The French immediately took this not as a legitimate killing, but as breaking rules of war. It also boiled down to "who fired first"—the French, or the British? The controversy led directly to the declaration of war.

Mary's oldest boy was a commander of forces, directly responsible for military tactics. And he enjoyed it. "I heard Bulletts whistle and believe me there was something charming in the sound," he wrote to his younger brother John Augustine.[5] If she was worried he would be in danger in the Royal Navy, she must have been frantic now. "Poor Mary Washington!" said Nancy Turner. "She had shorn him of one uniform only to see him don another."[6] She would have considered this a disobedient son, a betrayal of his earlier promise. But years had passed since then, and George was no longer a boy, but a man who managed his own plantation, away from his mother. She was mad, and, in Chernow's words, "hellbent upon preventing George" from reenlisting.[7] But it was a good lesson for him when he resigned after his failure. That meant he learned.

Mary soon found out he wasn't going to stay put.

In 1755, Major General Edward Braddock, a Scotsman and capable leader and veteran, arrived in Alexandria, Virginia, with orders to expel French forces from the Ohio Valley. George quickly introduced and acquainted himself with the general, and volunteered as an officer of the expedition.

Mary would have none of it.

She rode to Mount Vernon to make her case. Just as she had done years earlier, she asked that he stay. Now he had a plantation; that was more important than some distant land. He was an adult, yes, but that only put his life in more danger, not less. "Oh, this fighting and killing!" she yelled. That did not work. So she invoked religion, and evoked God Almighty, whose destiny for George was to stay at home. "God is our sure trust. To Him I commend you."

Without missing a beat, George was said to have replied, "The God to whom you commend me, Madam, when I set out on a more perilous errand, defended me from all harm, and I trust he will do so now."[8]

(Conveniently, the more hagiographical biography, *The Story of Mary Washington* by Mary Terhune, aka Marion Harland, skipped George's curt reply, instead explicitly saying that Mary approved of and blessed the venture.[9])

The talk might have gone on for some time—Mary apparently didn't take no for an answer kindly.

"My Mother [is] alarmd [*sic*] at the report of my intententions [*sic*] to attend your Fortunes," he wrote to Robert Orme, an aide to Braddock, in early April of 1755. His mother had prevented George from meeting with Orme, tarnishing a good impression of his military duties. It was humiliating, frankly, that he must miss such an important meeting. The Mount Vernon plantation was only six miles from downtown Alexandria, no considerable distance by any means on foot or horse. And he could not leave. "I find myself much embarrassd [*sic*] with my Affairs; having no person in whom I can confide, to entrust the management," he wrote.[10]

Mount Vernon had to be managed in his absence, and leaving affairs in disarray certainly would have vindicated Mary. As she was wont to say, "Ah, George had better have stayed at home and cultivated his farm."[11] Better to make a profit at home on the safe plantation of Virginia than to travel over mountains and rivers in Ohio, with the dreaded French and Indians wanting to put a sword or arrow in every redcoat they saw. It was clear to Mary which decision was better for her child. The responsibility of Mount Vernon ultimately fell on his brother John Augustine, but the argument between mother and son, again, was a thorn in his side.

Washington was nevertheless accepted to the Braddock Expedition as a colonel, and this time, no amount of begging or pleading would stop him. Mary, defeated in swaying her disobedient son, went back to Ferry Farm, alone. The expedition departed on April 16, 1755, from Alexandria.[12] Two weeks later, from Winchester, Virginia, George wrote to his mother that he was "very happy in the Generals [sic] Family, being treated with a complaisant Freedom which is quite agreeable." Perhaps Mary's insistence that his service in the Royal Navy would limit his "freedom" still stung George—perhaps this was a little gnawing back at his mother, the first of many times that this adult would disagree.

He had, he continued, "no reason to doubt the satisfaction I hoped for, in making the Campaigne."[13]

The gnaw became a bite.

IN EARLY JUNE, ONLY TWO MONTHS AFTER THEY DEPARTED VIRGINIA, MARY sent a letter to George, requesting, of all things, butter and a Dutch servant, referring to the wave of German immigrants into the colony. George in all likelihood would have been dumbfounded. Here he was, in the newly built Fort Cumberland, Maryland, over one hundred miles north, on an expedition to retake the Ohio, and she asks for butter and a servant! "We are quite out of that part of the Country where either are to be had," he replied to her. In fact, it was an impossible task, whether he wanted to deliver it or not. "Butter cannot be had here to supply the wants of the army," he pithily wrote.

He reminded her of what she agreed upon, that she would stay at Mount Vernon during his absence, which would allow his brother John to assist her. She probably did not visit, but nevertheless, George was diplomatic in reminding her: "I hope you will spend the chief part of your time at Mount Vernon as you have proposed to do where I am certain every thing will be orderd as much for your satisfaction as possible."[14] George wanted to take care of her.

SCANT LETTERS ARRIVED FROM EITHER OF THE TWO, AND NO RECORD OF any update exists until well past the defeat of the Braddock Expedition.

On July 9, 1755, near what is now known as Braddock's Field in Braddock, Pennsylvania, nearly two hundred miles northwest of Ferry Farm, and far away from motherly protection, General Braddock and his forces neared their goal of French Fort Du Quesne. The fort is located where the Allegheny and Monongahela Rivers meet to form the Ohio River. Suddenly, a surprise raid attacked and absolutely slaughtered the British forces. The surprise took any numerical advantage away from the British. "The dead," wrote David Humphreys in Washington's authorized biography, "the dying—the groans—lamentations—and crys [*sic*] along the Road of the wounded for help . . . were enough to pierce a heart of adamant."[15]

General Braddock himself was wounded and died four days later. George was also the target of many musket balls, though luckily for the country, none of them struck. Hundreds of the British were killed, and even more wounded, whereas the French and Indians had barely any casualties. It was a total and complete annihilation.

The news spread quickly, and Mary, upon hearing of the rumors, would have thought it the end of her son. A story spread that George was among the victims, having given brave and honorable last words before arrows and swords fatally pierced his body.[16] It was not an outlandish rumor, either. Most officers were killed, and George, she thought, reckless for such abstract things as *honor* and *valor,* would have been one of them. Any consolation that her boy was alive was futile—the Braddock Expedition was dead.

George would have sensed this worry. The same day that he wrote the governor, he also wrote his mother, just as much a priority, to get her the news. The letter, lengthy for a letter of his to her, was detailed, as it provided a firsthand account similar to his letter to Governor Dinwiddie:[17]

Honour'd Madm

As I doubt not but you have heard of our defeat, and perhaps have had it represented in a worse light (if possible) than it deserves; I have taken this earliest oppertunity to give you some acct of the Engagement, as it happen'd within 7 miles of the French Fort on Wednesday the 9th Inst.

We Marchd onto that place witht any considerable loss, havg only now and then a stragler pickd up by the French Scoutg Indns. When we came there, we were attackd by a body of French and Indns whose number (I am persuaded) did not exceed 300 Men; our's consisted of abt 1,300 well armd Troops; chiefly Regular Soldiers, who were struck with such a panick, that they behavd with more cowardice than it is possible to conceive; The Officers behav'd Gallantly in order to encourage their Men, for which they sufferd greatly; there being near 60 killd and wounded; a large proportion out of the number we had! The Virginia Troops shewd a good deal of Bravery, & were near all killd; for I believe out of 3 Companys that were there, their is scarce 30 Men left alive; Capt. Peyrouny & all his Officer's down to a Corporal was killd; Capt. Polson shard near as hard a Fate, for only one of his was left: In short the dastardly behaviour of thos[e] they call regular's, exposd all other's that were inclind to do their duty to almost certain death; and at last, in dispight of all the efforts of the Officer's to the Contrary, they broke, and run as Sheep pursued by dogs; and it was impossible to rally them. The Genl was wounded; of wch he died 3 Days after; Sir Peter Halket was killd in the Field: where died many other brave Officer's; I luckily escapd witht a wound, tho' I had four Bullets through my Coat, and two Horses shot under me; Captns Orme & Morris two of the Aids de Camps, were wounded early in the Engagemt which renderd the duty harder upon me, as I was the only person then left to distribute the Genls Orders, which I was scarcely able to do, as I was not half recovered from a violent illness that had confin'd me to my Bed, and a Waggon, for above 10 Days; I am still in a weak and Feeble condn which induces me to halt here 2 or 3 Days in hopes of recovg a little Strength, to enable me to proceed homewards; [f]rom whence, I fear I shall not be able to stir till towards Sepr, so that I shall not have the pleasure of seeing you till then, unless it be in Fairfax; please to give my love [to] Mr Lewis and my Sister, & Compts to Mr Jackson and all other F[rien]ds that enquire after me. I am Hond Madm Yr most Dutiful Son

G. W.

An interesting observation here: the inclusion of his personal danger. In a similar status report to the governor, George described the near-fatal encounter, and then scratched it out, deeming it unprofessional. Perhaps an intimate touch was considered fluff for the governor, and not so for his mother. Still, when reading the letter, and how disastrous it was, how he was nearly killed . . . Mary wouldn't have wanted to read that sentence. But he was safe, and disappointed in his illness that prevented him from seeing her, Betty, Fielding, all others, and trusted her word to update all as soon as possible. She was still his mother, and he knew that she cared for him as any mother would.

George returned to Mount Vernon within three weeks, on July 26.[18]

Mary must have been immensely relieved. Even Joseph Ball, Mary's half brother in England who so long ago helped solidify her decision to keep George away from the navy, applauded him: "It is a Sensible Pleasure to me to hear that you have behaved yourself with such a Martial Spirit in all your Engagements with the French Nigh Ohio. Go on as you have begun; and God prosper you. We have heard of General Bradock's Defeat. Every Body Blames his Rash Conduct. Every body Commends the Courage of the Virginians and Carolina men: which is very Agreable to me."[19] Joseph understood nearly a decade had passed since that pivotal (or decisive) letter to his sister, and understood that George was no longer a child.

Mary did not. She couldn't.

She wanted him to stay; she wanted him to be by her side.

He couldn't.

Despite the horrors and failures of the Braddock Expedition, George was commended as a hero, and promoted to commander of the newly formed and unified and combined Virginia forces, the highest military position of the state. "Commander-in-chief . . . twenty-three and a half years old," wrote Douglas Freeman.[20] The success was unbelievable. But there was still work in the Ohio—the French of course did not retreat, and the war was still raging—and no one seemed more suitable for the role than the young man himself.

But with the fame and success came a new batch of worries from his mother.

Mary, again, begged him not to go. Not to return to Ohio. He would definitely be killed or worse if he went back, she thought. He had nearly been killed before, as simply an aide to Braddock. On August 14, 1755, before his appointment, George wrote his mother a simple letter from Mount Vernon: "Honored Madam, If it is in my power to avoid going to the Ohio again, I shall, but if the Command is press'd upon me by the genl voice of the Country, and offerd upon such terms as can't be objected against, it woud reflect dishonour upon me to refuse it; and that I am sure must, or ought, to give you greater cause of uneasiness than my going in an honourable Comd; for upon no other terms I will accept of it, at present I have no proposals, made to me nor have any advice of such an intention except, End, from private hands. I am Dr Mm &c."[21]

The letter's original draft had scratched off "eternal" before "dishonour," showing not a simple dishonor for him but a damning one if he refused and accepted his mother's pleas again.

There was no eternal disgrace that day, for hours later, that same day he wrote Mary, he received word of his appointment.[22]

THE REMAINDER OF THE FRENCH AND INDIAN WAR UNDER THE GENERAL command of Governor William Shirley of Massachusetts, who succeeded Braddock, resulted in mixed results for the British cause. Colonel Washington of the Virginia Regiment led his troops in a series of battles, most impressively the Forbes Expedition in 1758, a rousing success for the British with the attack on Fort Du Quesne, which the French burned down rather than allow capture.

George tried to keep up correspondence with his mother, whose fear for her son's life—or, maybe, fear for her own comfort without her son—never abated. Some correspondences were minute and bland of character. At one point, he asked his mother to buy material for clothes, which otherwise were delayed from Britain, causing a headache for George, as winter was fast approaching and he needed clothes for his numerous slaves. "Therefore [I] beg the favour of you to choose me about 250 yds Oznbitgs 200 yds of cotton 35 pr Plad Hoes and as much thread as is necessary in Mr [Fielding] Lewis' Store if he has them if not in Mr [Robert] Jackson's and send them up by [John Alton, a servant] who comes down with a Tumbler for that purpose." In the same letter, he relieves her of his brother Charles's fears of impending marriage.[23] Also in 1757, a few months earlier, he visited Mary and gave her 5 pounds.[24]

Through a series of letters to Governor Dinwiddie, George led and commanded: strictly, emphasizing morality and discouraging profanity, with an infraction swiftly punished. He believed the "protection of Providence" kept his thousand-strong regiment safe.[25] In short, he was orderly, calling for obedience . . . much like his mother.

WITH HIS MILITARY SUCCESS IN THE OHIO, WASHINGTON WENT HOME AGAIN to Mount Vernon, to stay. Soon after, he married a lovely young woman, widowed once, by the name of Martha Dandridge, in 1759, only a few months after resigning his military commission, to the shock of many.

Mary, the strong and loving mother, could not have been happier. But it wasn't for George's sake.

Characteristically filled with spelling and grammatical errors, her short letter to her brother in England in late July, 1759, read: "I have known a great Deal of trouble sinc [sic] I see you[,] thear [sic] was no end to my troble [sic] while George was in the Army butt [sic] he has now given it up."[26] A glaring omission is the pride that she may (or may not) have felt for her son. Instead, it's—again, characteristically—focused on herself, at least on the surface. It was "my trouble" not "his trouble," though this could be read as either self-centered or selfless, because Mary was careful to cloak her frustrations in faux-humble language. Perhaps "my trouble" was not her inconvenience at his leaving; perhaps, like a decade earlier, it was her worry about his safety. It's true that when George came home to Mount Vernon, he didn't just come home closer to her, but he came home to peace and quiet, away from war and savagery. Maybe Mary even believed that her concern was truly for George's safety, not for her own feelings.

But there was also no mention or hint of Martha, six months married to George at this point. No mention of his heroics or his fame or his honor. Such things did not matter to Mary. For a small error-filled note, it was surprisingly telling of her older personality. Washington biographer James Flexner wrote that this letter "epitomizes the lack of sympathy or understanding that had made George find Ferry Farm an unhappy place once his father had died."[27]

Whatever relationship they had started to unravel.

THE SEVEN YEARS' WAR IN EUROPE ENDED IN 1763 WITH THE DEFEAT OF THE French and Holy Roman Empires. The British and their allies had won. In North America, the French relinquished all of their possessions east of the Mississippi River to the British, expanding the land beyond any colonists' wildest dreams. It ended whatever hopes France had for colonial expansion in America, making the British Empire the dominant overlord in the New World.

For now.

THE HOME LIFE OF MARY WASHINGTON, AWAY FROM HER OLDEST SON, WAS something of a routine. "Method became, with her, almost a mania," according to historian Sara Pryor. She never appeared late to church; she had her morning, noon, and evening plantation bells echo through the area exactly on time, so much so that "neighbors set their watches by" them. It was a strict observance of time that was clearly passed on to her son, which carried through into his later life, even as president, who was noted to be faultlessly punctual.

One neighbor who knew her reported that "Mrs. Washington never failed to receive visitors with a smiling, cordial welcome." However, "they were never asked twice to stay, and she always speeded the parting guest by affording every facility in her power." Pryor did suggest it was simply the manners of a host not to discomfort her guests by awkwardly forcing them to stay.[28] The quantity of food that was provided to guests—which came directly from the self-sufficient farm—was "incredible,"[29] especially the meat and bread. Robert Beverley, writing in *The History and Present State of Virginia* from 1705, described colonial Virginian food as "regaled with the most delicious fruits, which, without art, they have in great variety and perfection. . . . A kitchen garden don't thrive better or faster in any part of the universe than there." Decades later, the growth and production of meat, fruit, and grain only increased in quality and quantity from the fertile Virginian land like Ferry Farm.[30]

When not managing the plantation or the children, Mary slept in the rear hall chamber, which contained a bed, a cabinet, a tea table, four chairs, two windows and accompanying window hangings, a trunk, and other miscellaneous items. A fireplace kept her warm during the brutal winters. The 1743 appraisal of the house valued all items in the room to be worth over 19 pounds, a hefty amount.[31]

One relationship in her life that early biographers ignored was Mary's connection to the slaves. Who they were or where they came from made little matter; their purpose was the smooth management of her land, much like her father's and her father's father's time before.[32] She had multiple slaves throughout Ferry Farm, and in fact became a slaveholder with her father's death at the early age of three. Perhaps it was a formidable instrument to her later years. "From her earliest memories, she was the owner of other human beings," said Michelle Hamilton, historian and manager of the Mary Washington House in Fredericksburg, near Ferry Farm. "Her mother and then her half sister Elizabeth would have instructed Mary on how to be a mistress, how to command respect, how to manage others. This would have developed in Mary her iron will that we see in her as an adult."[33] Slaves were no exception to this "iron will," and receiving these other human beings at such a young age may have diluted the moral importance of them. To Mary at age three, they were servants, maybe even people to play with. But they did what she said.

"In her dealings with servants, she was strict," said Douglas Freeman. "They must follow a definitive round of work. Her bidding must be their law."[34] But there is no contemporary record of her slave keeping or slave management. The lack of note, especially living so close to the port, is good proof of her being a "by the book" owner. If you were strict or excessive in abuse, it was talked about.[35] Betty, her only surviving daughter, also had no qualms of historical note with her servants, except for the usual runaway once or twice. If Mary taught Betty to be cruel like herself, there would've likely been a record of it.[36]

Tradition said that one overseer, after doing some unknown task his way and not Mary's, was scolded. "Madam," he said to the mistress, "in my judgment the work has been done to better advantage than if I had followed your directions."

"And pray," she replied, angrily, "who gave you the right to exercise any judgment in the matter? I command you, sir; there is nothing left for you but to obey."[37]

The relationship between slave and mistress came to a head toward the end of 1750 when the slave Harry murdered Tame, another of Mary's slaves. He went to trial, which concluded on September 7 of that year. The King George County court records said thus:

> *Harry a negro slave belonging to Mary Washington of this county—being brought [out] of gaol [jail] and upon an Inditement arranged at the Bar for the murder of Negro Tame a man slave also belonging to the said Mary Washington Plead not guilty. The Court upon hearing the witnesses and duly considering the case of opinion that the sd Negro Harry is guilty of the said murder said in the said indictment and [thereupon] do order that the said Harry do return to the Gaol [jail] from which he came, and the Sheriaf on Wednesday the 10ᵗʰ day of October next, take him thence and carry him to the place of execution and then and there to hang him by the neck until he be dead.*

Harry was executed on schedule, and Mary, now short two slaves, was reimbursed 35 pounds from the county for her troubles and the loss of property.[38] The murder of slaves or slaves committing murder was not an uncommon occurrence, noted the *Virginia Gazette* several times, often with a flat, distant tone denoting no more than their monetary worth. Right below many announcements were advertisements of slaves for sale at varying prices, with the same businesslike tone.[39]

Less than a year after the murders, other slaves—in this case, indentured servants—stole property. The story went that George was swimming in the Rappahannock, and two servants, Ann Carroll and Mary McDaniel, were accused of stealing his clothes. Both were tried. Carroll was "discharged," according to court records, and McDaniel was found "guilty of petty Larceny," and immediately was to be taken "to the Whipping post & Inflict fifteen lashes on her bare back."[40]

These two incidents, though only two, may have vindicated how others saw a woman leading a family and owning land. Laura J. Galke wrote, "The Fredericksburg community likely viewed such tragic events as proof of an unsupervised woman's inability to manage a plantation."[41] Indeed, many women remarried for just such a reason: they didn't know how to manage a farm.

Whisperings and rumors of failure at Fredericksburg probably did not have much impact on the isolated life of Ferry Farm. The tobacco transports at the Rappahannock came and went without any regard to Mary's own schedule. In a July 1760 letter to her brother Joseph in England, she wrote, defensively, "You seem to blame me for not writeing to you butt I doe ashure you that it is Note for want for a very great regard for you and the family, butt as I don't ship tobacco the Captains never call on me, soe that I never know when tha com or when tha goe."[42] The letter, riddled with misspellings, showed a woman who had no education or had no need to correct her writing. Even for an age with no standardized spelling, Mary wrote in an uneducated, unmannered way. She had her life settled at the farm, no need to impress beyond necessity.

Isolated she may have been, but Mary understood her social status as both a widow and a Washington near Fredericksburg. She may have dabbled in embroidery or tambouring, which would have solicited visitors to admire her skills and works. One tambour hook discovered on Ferry Farm, placed in the cellar between 1741 and 1760, may have been used by either Mary or Betty, or both, and contained a bone handle and steel hook, and was heavily decorated with designs.[43] Archaeological findings of 1740s- and 1750s-styled teapot stands, silver teaspoons, and ceramic tea cups and saucers indicate a social life to some degree, even if Mary insisted on independence. "Mary's frequent purchases of English-made tea wares throughout her life also highlight her determined efforts to communicate her family's refinement and fashionability using sensibly priced and unassuming ware types," theorized Laura Galke. She continued, saying that it is without a doubt—based on the discovery of a teaspoon with the inscription of "BW"—that Mary also placed these social customs on her children, especially Betty.[44] Her uncle Joseph Ball even provided silverware for the young Betty as far back as 1749; "I have sent you by your brother Major [Lawrence] Washington a Tea Chest," he wrote to her. "And in it Six Silver Spoons, and Strainer, and Tongs of the same." Even authentic British green tea was given to her. "As soon as you get your chest you may sit down and drink a Dish of Tea," he promised.[45] This indicated a clear interest in keeping her reputation, despite whatever financial or isolationist manners she had.

Mary's other children also matured and grew up, though with not nearly the attention placed on George by both Mary and historians. Elizabeth, nicknamed Betty, Mary's first daughter, married Fielding Lewis on May 7, 1750. Betty, Mary's only surviving daughter, would have bonded with her mother in typical eighteenth-century activities for women, including sewing and embroidery.[46] George genuinely enjoyed her company. They got along so well and were so much alike, it had been passed on, that if she wore his clothes and bundled up her hair under a hat, they looked identical.[47] Fielding, with a lazy left eye that he wasn't ashamed of showing in portraits, was a well-known plantation owner and Fredericksburg retailer and friend of the Washingtons. His first marriage to Catherine Washington, the daughter of George's cousin John, lasted a little over three years, until her death.

The relationship of Fielding Lewis and Mary Washington went back years before he married her daughter, Betty. Fielding's first child, John, was born on June 22, 1747, and among the many godparents to the little boy was Mary herself.[48] Her religious ferocity would have made this as important a role as raising her own children.

Samuel Washington, born only two years after George, lived in Chotank. He married up to five times, a year apart from each of his wives' deaths. He was a sore sibling to George on his lifestyle. "In Gods name how did my Brothr Saml contrive to get himself so enormously in debt? Was it by purchases? By misfortunes? or shear indolence & inattention to business?" Washington wrote in January 1783, far in the future.[49] Samuel settled in Berkeley County (now in West Virginia), and died in 1781, forty-seven years old, having become a colonel in the Continental Army.

John Augustine Washington was George's favorite sibling. He "was the intimate companion of my youth and the most affectionate friend of my ripened age," George wrote to Henry Knox many, many decades later, soon after John's death.[50] John married Hannah Bushrod and went to Westmoreland County.[51] Evidence suggests a nice relationship with his mother, Mary, who affectionately called him "Johnne" and wished love to his family. An undated letter from her opened, "I am very glad to hear you and all the family is well," giving a sense of familiarity to her son's family.[52]

Only scant facts can be learned about the youngest Washington, Charles. He outlived all the Washingtons except George, dying only three months before his oldest brother. Charles's children, especially George Augustine, were a delight to the elder George.[53] Charles lived from 1760 to 1780 just steps from his mother, so he must have felt her presence at times. He had married a cousin, Mildred Thornton, the daughter of a prominent but not-as-prominent family in the region, and had all his children in the house. He was a vestryman of Saint George's Church. Afterward, he settled in Berkeley County, building his home Happy Retreat, in the present and eponymous Charles Town.

There is little record of Mary's treatment of her younger sons. They were neither the oldest nor the heir; perhaps they didn't receive as harsh care or want from Mary. They weren't as successful as George, even before the Revolution, and they weren't as sought after as he was. Samuel took part in the Revolutionary War, so perhaps her fear of losing not only her son George but her son Samuel weighed heavily on her. Her son Charles, as her youngest, may have been her favorite—the youngest child typically is, stereotypes say. He was supposedly an alcoholic (appropriate, then, that his home was converted into an unnamed tavern in 1792 and remained so into the 1830s). Michelle Hamilton called him the "black sheep" of the family, though Katie King, manager of the Rising Sun Tavern (Charles's former home), was hesitant to damn him so. "Charles couldn't have done so bad for himself, because he did manage to sustain his property, and then moves out to Happy Retreat." It wasn't debilitating so much that he "lost fortune."[54] He also probably had the best relationship with his brother-in-law, Fielding, even in business partnerships. Such a close-knit relationship could have hit a soft spot for Mary.

None so occupied her time and mind as George, though.

THE 1760S MOVED BY QUIETLY AND QUICKLY FOR MARY AT FERRY FARM. THE thirteen colonies, and Virginia in particular, were undergoing a boost of population, growth, and prosperity. "It was high noon in the Golden Age!" wrote Sara Pryor. "Life was far more elegant and luxurious than it was even fifteen years before."[55]

George had a life of his own, as did Mary's other children. They all left her at Ferry Farm. No correspondence between her and her eldest has been discovered from this decade. George still saw her, although extremely rarely after 1760. He visited for the night at his mother's in early January 1760. It was snowing, though the wind, strong to the southwest, was "not Cold," he noted. He stayed the night and the following morning, and overnight the snow turned to a mix of sleet and rain.[56] Six months later, he returned to Fredericksburg, and received 1 pound, 3 shillings, and 9 pence from a John Gist for some unspecified smithwork for his mother, but a few transactions and his diaries provide the only information.[57] He stayed at Fredericksburg during the Christmas of 1763, and again visited his mother in March 1769, where he gave her 3 pounds cash. He returned to see her in early November of that year.[58]

In the 1760s, she went from a woman in her fifties to one in her sixties, a major shift in a demographic and a milestone in its own right. She was a grandmother to many small children, none of which came from George. And she continued to stay at Ferry Farm as nearby Fredericksburg bustled with excitement and expansion. She presumably kept contact with the city, keeping in mind her social status as a Washington, not to mention her additional fame being the mother to George. In December of 1766 and 1767 there were musical concerts and balls, both hosted by John Schneider for a small price of 7 shillings and 6 pence. Violins, tenor, bass, flutes, horns, and harpsichords provided ample opportunity to dance and entertain. Mary, being a grand dancer—certainly, her son inherited that trait—could have easily passed these evenings as the mature madam dancing with men of all ages, young and old.[59]

But she probably preferred the solitude of home, being alone but not lonely. "Mary Washington was too busy to be lonely," opined historian Virginia Carmichael. Older she may have been, but energetic she was, and she kept her plantation at Ferry Farm in working order.[60] It helped both her and her farm that Fielding Lewis and his wife, Mary's own Betty, were nearby.

SO HER LIFE PASSED IN HOME FARM. THERE, "THE SEASONS MOVED ROUND into place, each as full as it could hold of duties and responsibilities, but full, too, of the content that comes only after strife."[61] Winters were brutal but the fireplaces provided warmth. The summers, too, were brutal: hot, in which case the trees provided shade. One of her most prized possessions, a fan, also did the job of keeping her cool. It was an import, from Asia, probably China. Its boxed case, also well used, was designed with swastikas, a holy symbol for some East Asian religions. To Mary, it was nothing but an eye-pleasing decoration, but one that was fancy enough for her status and practical enough to use; she kept it.[62]

Those intervening months while the world moved on were the best parts of her retirement. Ferry Farm was where she was to stay, if she wanted.

But her son thought otherwise.

THE YEAR 1770 WAS AN ACTIVE YEAR FOR MARY. ON JUNE 25 AND 26, JULY 31, and August 1 and 9, her eldest son dined and visited with her, before going elsewhere on to other business with his brother-in-law or others.[63] Many times, when seeing her, he gave her money or a gift as she requested, no more than 40 pounds at a time.[64]

Clearly, he wanted to talk to her about something.

Specifically, moving out of Ferry Farm—Home Farm, as it was known. The previous decade he had tried to get her to move out, but she steadfastly refused.[65]

Mary had continuously lived at this farm since its purchase by her husband in 1738. Though the farm was legally George's, he clearly did not mind her occupying the space. By 1770, thirty-two years had passed here, comprising a majority of her life.

In September of 1771, George surveyed half of Ferry Farm, noting the size, acreage, landmarks, roads, neighbors, boundaries, and other points of interest. He stayed a couple of days with Mary before heading out again to Fredericksburg, seeing Fielding Lewis to discuss the selling of the lands.[66] No original survey exists. It was, however, reconstructed based on existing notes, in *The George Washington Atlas* in 1932, published by the government-owned and historically keen George Washington Bicentennial Commission. The survey marks fences, such as the cornfield fence about halfway through the property, running north and then northwest. To the right was, George noted, "some pretty good land." The rest of the land has no explicit praise, perhaps a note of its decline in quality and production. Thirty-plus years on the same land did deplete nutrients in the soil.

Here, over the years, formed Rappahannock's ferry lane, which was unusually close to the plantation house. As Fredericksburg became more and more popular, so too did the ferries and its cargo of passengers, passing close by to the elder Mary.[67] It was time for her to move on.

A scary bout of illness in December of 1771, treated by Hugh Mercer, probably influenced the decision more than her pleading son. From December 20 to December 31, Mercer tended to Mary for some disease. He bled her; he gave her pills for vomiting; he gave her enemas, some "purging pills," and other remedies. The cost was, in total, 2 pounds, 18 shillings, and 6 pence.[68] (Genevieve Bugay at the Hugh Mercer Apothecary Shop in Fredericksburg believed it was the flu: a curable disease now, a scarier ailment three hundred years ago.) It is unknown why Mary went to Mercer and not her usual doctor, Charles Mortimer, but Mercer had helped her son Charles earlier that year on several occasions. In June, he helped Charles with "a Pectoral Electuary," indicating a chest muscle issue. In early October, he treated Charles for an unknown ailment by "purging" him, which could have been bleeding, vomiting, or laxatives. All visits from Mercer—eleven in total—perhaps provided a sense of comfort for Mary. Mercer had met George Washington in the French and Indian War, as part of the Pennsylvania regiment, and, when things settled down, took the surgeon's path. He had also gone to Mount Vernon to assist in Patsy Custis's epilepsy treatment. There was a Washington connection here, certainly, but why she went to Mercer for this one time is ultimately speculation.[69]

All her children, of course, had lives of their own. "Mary's out there by herself." It wasn't necessary for her to be there. Betty, too, helpful as she was, was unreliable for travel back and forth, having just given birth to her eleventh, her final, child, Howell.[70]

Further, to make it more apparent how secluded she was, in late January of 1772, a blizzard of two to three feet of snow hit the region. Now called the Washington-Jefferson Blizzard, the weather isolated Mary across the Rappahannock.[71] January 27 was "dreadfully bad," according to Washington's diary, as constant snow and wind pummeled the region.[72] What chance he had to leave his Mount Vernon house was nearly impossible without the greatest strength or difficulty to himself and his horse. Mary, then, was in even worse shape. If another storm were to come in her years at Ferry Farm, at an even more advanced age, she could suffer from the cold and/or deprivation.

Illness, lack of proximity to children, and isolating weather: for a lady her age at that time, it was clearly a bad idea to be alone.

By the spring of that year, she departed from Ferry Farm, and moved into a two-story house close to her daughter Betty. She had been offered a chance to live with Betty and Fielding, as was customary. It was normal for a widowed mother to be living under the roof of a child. All her property was appraised, and she was given an allowance by George, totaling 215 pounds.[73] Her evaluated items at her Ferry Farm plantation included seventeen hogs, valued at 5 pounds, 13 shillings, and 4 pence total; twenty-six shoats (young pigs) and older pigs at 5 pounds, 4 shillings; six sheep "very indifferent" at 4 pounds; three oxen, "one old" at 11 pounds; and other livestock. Overall, her total property at both Ferry Farm and elsewhere was appraised at 215 pounds, 11 shillings, and 8 pence.[74]

George had purchased a house on two lots on Charles Street, Fredericksburg, slightly on the outskirts of the city itself. Total payment was to Michael Robinson, the previous owner, for 275 pounds, the first payment of 75 pounds delivered in May of 1771. The sum was quite a bit, requiring borrowing money from Fielding Lewis. Everything was settled by the end of 1772, but Mary had moved in that previous May, per an agreement with Robinson and Washington.[75] In November of that year, George Washington formally placed Ferry Farm and a nearby plantation up for sale.[76]

THE CITY HAD CHANGED GREATLY SINCE MARY, AUGUSTINE, AND THE young children had moved there in 1738. It had become the tenth-largest port city in the colonies, with over two hundred ships passing by each week, an average of twenty to thirty per day.[77] Locals had become more prosperous as well. On Caroline Street, very much the "main street" of Fredericksburg, was the Indian Queen Tavern, built in 1771 and expanded in 1790 to include a billiards hall and stable. The spot proved popular, and many locals unknown and famous paid a visit. It would remain open until 1832, when a fire destroyed everything.

In 1752, on Princess Anne Street, was organized the Masonic society of Fredericksburg Lodge #4. Fielding Lewis, George Washington, William Woodford, George Weedon, Hugh Mercer, and so many other gentry joined this prestigious institution. A nearby cemetery, starting in 1784, offered a refuge for the Lodge's dead.

The 1750s also saw the operation of Hunter's Iron Works, just south of the Rappahannock River, founded by James Hunter. The plantation of Chatham, also built in 1750 by the influential William Fitzhugh, sat on the east side of the Rappahannock, overlooking the growing town.

Other buildings and shops opened in and around the town of Fredericksburg: Fielding Lewis took ownership of his father's store on Caroline and Lewis Streets in 1749; meanwhile, by the time of Mary's move, construction on the future "Kenmore" plantation home had commenced and would not be completed until the Lewises' move in the autumn of 1775. At least seventy servants and slaves built it, a major undertaking for four years that included intricate plaster work and material. The plantation, which was never called Kenmore (or anything, which was unusual) until the nineteenth century, would tower over all of the town, a major family seat that rivaled the Fitzhughs' Chatham Manor across the Rappahannock.[78]

ALSO FOR SALE IN 1772, IN AUGUST, AFTER MARY MOVED FROM FERRY FARM, but before the land was for sale, was another tract of land near Fredericksburg. About three hundred acres was for sale in accordance with the will of its previous owner Roger Dixon, who had recently died. For a to-be-negotiated amount, the advertisement noted that it was "adjoining the Town, several unimproved lots, the Ferry opposite to Mrs. Washington's, and a Lot, with Houses."

Three articles down, in the same column of the *Virginia Gazette*, with as much normalcy as a note about selling land, was the notice of a runaway slave from Tappahannock, about fifty miles to the southeast—a "well made, light colored Mulatto Wench named PHEBE," as it said. Twenty-two or so years old and about five feet four, Phebe, it was suspected, ran away months prior after being punished by her master for "a Propensity of Pleasures in the Night." Archibald Ritchie, the owner, offered 40 shillings for her whereabouts.[79]

Business as usual for the times.

MARY'S NEW HOUSE WAS TWO STORIES, SITUATED ON CHARLES STREET OF Fredericksburg, consisting of a small room and bedroom and a small attic. George stayed up in the attic, and at over six feet tall would have needed to duck and crouch down. It was new, built sometime in the mid-1700s, and contained a basement of brick and stone and wood, the same material of the walls. It was—and is, even with the additions through the centuries—a modest house. Facing northeast, the property also contained a small separate kitchen to the east, which had centuries later been expanded into the main building. It was within literal shouting distance of her daughter Betty.

Mary hated it. She hated the move, feeling it was forced upon her by her son. She had been comfortable at Ferry Farm. "There is a pain in moving out of an old home, bag and baggage, that stands all by itself in the catalogue of human agonies, a pain that this woman, of all people, must have felt to her inmost being," wrote author Nancy Turner in the early twentieth century.[80]

Ferry Farm plantation passed from history, and from the Washingtons, in 1774, when Dr. Hugh Mercer, physician, neighbor in Fredericksburg, fellow Fredericksburg Masonic member, and friend to George, bought it all for 2,000 pounds, to be paid in five annual installments.

Moving away from her home, even a mile away, could prove no small feat. But in Fredericksburg, despite being closer to her daughter, she said she was living "in great want."[81] She even borrowed food and money from a neighbor and son's friend, Edward Jones, who was charged of overseeing Mary's needs and, previously, that of Ferry Farm's. In late 1772 he noted that he gave her lamb, shoats, and corn.[82] She was being a headache not just to George, evidently. George, after Mary's death, later wrote to Betty that she "took every thing she wanted from the plantation for the support of her family, horses &ca."[83] There is even a tradition that Mary would drink only water from Ferry Farm, she so refused to settle down.[84]

Yet, there was a sense of humility here. Few artifacts remain of hers, but those that do are telling. A small, tinted mirror, now glazed over and unviewable, decorated a wall. It's simple, not ornate. Another, a teapot, was imported from China, with rich decorative designs. One shows a man riding a bull, a curious but fun decoration.

She would stay here for the rest of her life. Perhaps this was as intended by George, to have her settle down for her last years.

Finally.

Chapter 9

Before the Revolution

THE COLONIES AND GREAT BRITAIN

1763–1775

"If this be treason, make the most of it."

T he New World was beginning to tremble . . . and soon, within four short years, a political, social, and religious quake would shatter the colonies, making them unrecognizable to the earliest settlers. From simple and separate entities a century earlier, they would soon become a new unified nation with a new constitution, and a new system of government that marked the beginning of the end of an empire.

Life may have gone on as normal for Mary Ball Washington at Ferry Farm in the 1760s, and things may have been settled for her; but as she managed the acres of land and her servants and slaves, and as she pestered her son for money, material, anything, the colonists whispered in hushed angry tones at the increasingly demanding laws and restrictions from across the ocean. This was not something new—it had gone on for decades—however, in recent years, tensions had become increasingly strained between England and the colonies. It was inevitable, like a child rebelling against his or her parents. By the 1770s, that fight was about to rise to a crescendo.

THE THIRTEEN COLONIES OF AMERICA WERE NEAR-PERFECT EXAMPLES OF royal obedience ever since Jamestown was settled in 1607. They had fought wars for the Crown, had sweated and bled for imperial expansion. Colonists by legions volunteered in the militias and the armies. "God save the king!" was a proclamation of subservience to the king and queen, who, by the grace of God, ruled. George Washington himself had fought under the British Crown, as had his stepbrother, Lawrence.

Due to its distance, the colonies were granted relative independence from the inner workings of England's Parliament. Some governors and acting governors were effectively semi-autonomous rulers of their respective regions. Virginia, in a way, set a precedent for independent rule, as the House of Burgesses, of which many Washingtons and Balls were members (including George, in 1758), was an elected legislative body of local planters, lawyers, and clergymen, the first of its kind in English colonies.

To be sure, independence did not mean outright disloyalty to the Crown. And independence did not mean that no conflicts existed between the Mother Country and her children. Colonists were still colonists, whether they were on the other side of the world or not. Governors, especially of Virginia, were directly appointed by the king, and often never even visited the colonies, instead allowing the lieutenant governors to rule in their stead.[1] Yet the colonies still "cherished the most tender veneration for the mother country," wrote John Corry in 1809. In return, there would be the "protection of Britons, and [they] witnessed their valour with admiration."[2] It was a relationship certainly advantageous to both parties, but it was clear which one was in charge.

In law, the colonies were purely English. Common law was, well, common, affirmed by experts through centuries of thought. One man, Richard West, lawyer of the Board of Trade in London, said in 1720 that colonists and their lands were not subjected peoples and enjoyed the same rights as others. Likewise, all laws of England were to affect the laws of the colonies—that was the idea, at least.[3]

But that was in 1720—many things were to change in subsequent years.

THE CROWN CHANGED HEADS ON OCTOBER 25, 1760. THE PRINCE OF WALES, George William Frederick, of the House of Hanover, became king of Great Britain and Ireland on that Saturday, becoming George III, after the death of his grandfather, King George II, who had reigned for over thirty years. George III was twenty-two years old upon his ascent.

Unlike his grandfather and his great-grandfather, this new king was born and raised in England. His predecessors rarely if ever visited their kingdom, instead spending their days on the European continent. So when George III was announced as the new royal, the news was met with celebration both at home and abroad.[4] He even felt pride in it, as noted in his ascension speech to Parliament.[5] In 1766, Virginians were called to celebrate the sixth anniversary of his ascension to the throne at the governor's palace in Williamsburg. The late-October date was in lieu of the celebration of the king's birthday in June, a move that the *Virginia Gazette* noted was "a more agreeable and convenient season for the company to pay their compliments, and show their respect, to his Majesty, than the heats of summer."[6]

King George III was raised mostly by his mother, Augusta, Princess of Saxe-Gotha-Altenburg by birth and of Wales by marriage. His mother would often remind him to "be a king, George," no small task but a vague enough suggestion to make or break his reign.[7] How different from Mary Washington's determination to inhibit her son's rise.

He would reign until his death in January 1820, at the advanced age of eighty-one. But forty-four years of his kingdom would be troubled, as the colonies divorced him and the British Empire.

IRONICALLY, THE WAR THAT BROUGHT GEORGE III INCREASING POPULARITY and fame, the war that distinguished the colonies as vast lands worth keeping, more than just backwater frontiers, was the same war that began the unraveling of England's American empire. The French and Indian War was a catalyst. In 1763, when the war ended, the colonies became increasingly intolerant of England's rules.

In previous years, these abusive acts or laws were endurable. Colonists accepted England's demands during the French and Indian War. Some did more than others. "The colonists in Alexandria [Virginia] like Anglo Saxons everywhere," said William Carne in the mid-1800s, "had a profound respect for *law*. They submitted to these restrictions because they were accustomed to them, and as they were pressed by French power, and in constant dread of the Indians, who still lingered near, they expected British aid, and thought the profits, which England made by a monopoly of their trade, was a high price to pay."[8] The French and Indian War was still a war, one that the colonists weren't ready to fight alone.

But that would change.

King George III immediately made his thoughts known, calling the French and Indian War—and the greater Seven Years' War—a "bloody and expensive war," which shocked some of the ministers and members of Parliament, who were sure of English dominance. Men died in wars, and wars cost money and materiel, this was obvious—but the success brought so much more than what was lost, they thought.[9] George thought otherwise.

The war doubled Britain's debt from the decade prior, an increase of more than 70 million pounds. It was an impossible number. How would England repay that debt?

With the colonies' help, clearly. After all, it was their doing that got England into this mess, they reasoned in England. Prime Minister George Grenville and King George III would see to that.

IN 1763, THE SAME YEAR THE FRENCH AND INDIAN WAR ON AMERICAN SOIL and the Seven Years' War in Europe ended, King George III issued a proclamation forbidding all settlement of the newly acquired land. It coincidentally followed a rebellion of Indian tribes in the Great Lakes region, the Pontiac Rebellion, which resulted in a stalemate. The proclamation deemed that the large strip west of the Appalachian Mountains and east of the Mississippi was to be left to the Native Americans as a reserve to protect and calm them, lest they attack incoming English settlers. This voided any promise of land to those who wanted to expand westward.

There were colonists who lived there, they had just fought for the land, and the king—God save the king, indeed, from himself—proclaimed that they could not settle there. So much for the war, and so much for the Crown's interest in colonial settlements.

The proclamation itself did not last long and was inevitably pointless, as treaties with the Indian nations and with tribes in 1768 made the initial issue by the king unnecessary.

George Washington himself was affected, angry even, and several years later took charge of securing two hundred thousand acres—as was promised—for his troops at Fort Necessity. He called it a "cursory manner" in a letter to Governor Norborne Berkeley, Baron of Botetourt, but in private he must have been seething.[10]

THE REVENUE ACT OF 1762 WAS IMPOSED TO STOP ILLEGAL COLONIAL TRADE of molasses and other goods between so-called smugglers and the French, who circumvented customs agents. This spurred, over the next five years, act after act of more taxes and restrictions levied by Parliament on the colonies. British soldiers searched houses with impunity, infuriating the colonists.

In 1763, the Customs Act was similarly intended to increase enforcement against molasses smugglers. British naval warships often blockaded ports of trade, with ships of the line unyielding over colonial harbors.

In 1764, the Sugar Act, replacing the ineffective and earlier Molasses Act, placed a smaller tax on sugar and molasses to 3 pence per gallon, instead of the previous 6 pence. It was a noble intention to some, bitter medicine to others. Richard Jackson, a member of the House of Commons and correspondent of Benjamin Franklin, supported it, though he proposed an even smaller tax of 1.5 pence per gallon. Franklin, writing from Philadelphia, however, was cautious, saying, "If it is not finally found to hurt us, we shall grow contented with it; and as it will, if it hurts us, hurt you also, you will feel the Hurt and remedy it." He noted it was making "a great Stir among our Merchants, and much is said of the ill Effects."[11] But with the drop, which was still deemed unnecessary, came stricter penalties and enforcement previously unseen.[12]

In 1765 the infamous Stamp Act was imposed. This was one of the most notorious—and perhaps the most egregious—of all the acts in the 1760s against the colonists. "The most momentous act of the Grenville ministry is not mentioned in the correspondence between the King and his minister," wrote historian John Brooke. It was not outrageous to the high and mighty Britons.[13] The other acts were more or less annoyances in comparison. The Stamp Act was purely internal, the first of this series, making it a wholly colonial affair.

What protests there were from the colonies before its passage— and there were many—were unheeded. "We might as well have hindered the sun's setting," wrote Franklin in his autobiography.[14]

The disastrous effect of the Stamp Act could not be under-stated. It was immediate and outrageous. Writing in 1785, John Andrews, in his book of the Revolution, wrote that "this famous act has justly been considered as the prelude and occasion of all the subsequent storms. . . . Its arrival in America threw immedi-ately the whole continent into flames."[15]

This one called for stamp duty on newspapers and all legal documents, enforced by tax collectors. This included land grants, surveys, legal letters and documents, mortgages, birth certif-icates, and bank loans, all of which were to be taxed at over 3 shillings per page. Other documents like warrants and grants were also included. Attorney licenses received the heaviest tax, at 10 pounds. For the first time, organized opposition to the Stamp Act specifically, and England generally, was heard throughout the colonies.

It was all-encompassing, affecting both the well-to-do and the more modest of colonists. For surveys with smaller tracts of land, the tax added up. For example, "a modest 200 acre plot in Vir-

ginia, worth about six pounds Virginia currency, would have required stamps on the grant, the warrant to survey the land, and
the registration of the land. This amounted to three shillings,
three pence sterling, or five shillings, three pence Virginia currency, slightly more than 4 percent of the purchase price," wrote
Claire Priest.[16]

The tax was not necessarily the issue, though it certainly added
fuel to the fire between Parliament and the colonies. The tax was
ineffective and rage-inducing, but it was ultimately just about
money. The underlying issue was the colonists' identity and value
for the Crown and for Parliament. Who they were in law versus who they were in practice. The issue, in short, was that there
was no consent or consensus with the colonists. This was, again,
a wholly internal tax, and should have been decided by fellow
colonists, not some distant legislature. The Virginia House of Burgesses had pleaded with the king not to break this tradition, but
it was to no avail. The Stamp Act gave rise to the rallying cry "no
taxation without representation." This issue went back decades;
no colonist represented them in the House of Commons. Franklin
noted as such in 1754 in a letter to Governor William Shirley of
Massachusetts.[17]

George Washington was an outspoken critic of the Stamp Act,
saying it "engrosses the conversation of the speculative part of the
colonists, who look upon this unconstitutional method of Taxation
as a direful attack upon their Liberties." He continued in the letter, wondering "who is to suffer most in this event—the Merchant,
or the Planter." It was a multipage dispatch to the London-based
Robert Cary & Company disparaging the legislation, written by
an uncharacteristically furious Washington.[18]

To Francis Dandridge, a relative of Martha Washington's in
London, he wrote a similar letter that same day.[19]

It ignited a flame in the colonies, and as opposed as Washington was, there were others more provocative, more telling in their
intent. Patrick Henry, a fellow Virginian, was one such agitator, a
man Thomas Jefferson compared to Homer. "Tarquin and Caesar

each had his Brutus," Henry exclaimed to the House of Burgesses on March 30, 1765, "Charles the First his Cromwell, and George the Third . . ." Aghast shouts of "Treason!" echoed through the chamber, interrupting his speech. One delegate, George Johnston of Alexandria, stood next to Henry in support.[20]

Henry was said to have replied, "If this be treason, make the most of it!"[21]

THE STAMP ACT WAS REPEALED BY PARLIAMENT IN EARLY 1766, WITH THE caveat that Parliament still had the power to assert laws on the colonies, with or without permission. But it was too late. The repeal of the Stamp Act nonetheless left a bitter taste in the mouths of many colonists.

Up north, in Boston, Massachusetts, the center of ideas and events that sparked so much of the coming decade, was an organization called the Sons of Liberty. This group's idea of taxation without representation spread far and wide in the thirteen colonies. In Virginia, in May 1766, a group four hundred strong met at Hobbs Hole—currently known as Tappahannock, Virginia, with the Rappahannock River only a stone's throw away. The *Virginia Gazette*, with the subtitle "Open to ALL Parties, but Influenced by None," published this group's full-page announcement. It affirmed loyalty to King George III, but noted that they, in the interest of the people of America, would not follow the Stamp Act. "The Stamp-Act does absolutely direct the Property of the People to be taken from them without their Consent," it said, "expressed by their Representatives, and . . . in many Cases it deprives the British American Subjects of his Rights to Trial by Juries." These Sons of Liberty wanted no Virginian or colonist to give an inch to Parliament. Yielding even a little in the law would acknowledge that Britain had a right to tax without consent.[22]

Whatever tremors there were in 1763—if one couldn't feel them then—could certainly be sensed now.

JOIN, or DIE. Franklin's political drawing of a colonial snake with a severed body from a decade earlier took on a life of its own.

PARLIAMENT LEARNED NOTHING. UNDER THE CHANCELLOR OF THE EXCHE-
quer and former president of the Board of Trade Charles Townshend, it passed the Revenue Act of 1767, known as the Townshend
Act. This act imposed a new tax on glass, paints, paper, and tea. It
was another slap in the face of colonial independence. Even Benjamin Franklin, a man of intelligence and knowledge in philosophy
and political thought, was puzzled. "The Sovereignty of the King
is therefore easily understood," he wrote that year. "But nothing
is more common here than to talk of the *Sovereignty of Parliament*, and the *Sovereignty of this Nation* over the Colonies; a kind
of Sovereignty the Idea of which is not so clear, nor does it clearly
appear on what Foundations it is established."[23]

The outrage in the colonies—Boston, again, primarily, a hot
spot for revolutionary thought—was so great that in 1768, British
troops were ordered in to enforce the law. As with the Stamp Act,
the Revenue Act was partially repealed in April of 1770, except
for the 3 pence-a-pound tax on tea, which remained. "Even this
trifling impost kept alive the jealousy of the colonists, who denied
the supremacy of the British legislature."[24] It was clear Parliament had no desire to help or even support the colonists.

Back at home in Virginia, the House of Burgesses passed, in
secret, a measure directly and explicitly condemning the British
Parliament. Voted and passed on May 16, 1769, it stated in no uncertain terms "that the sole Right of imposing Taxes on the Inhabitants of this his Majesty's Colony and Dominion of Virginia,
is now, and ever hath been, legally and constitutionally vested in
the House of Burgesses . . . with the Consent of the Council, and
of his Majesty, the King of Great-Britain, or his Governor for the
Time being." Other resolves included the right of Virginians to
plead directly to the King, as well as the complaint of being tried
in England.[25] George Washington was among the signers.

Governor Berkeley dissolved the House of Burgesses immediately upon receiving word of the measure. In response, the dissolved
members went to Anthony Hay's Raleigh Tavern in Williamsburg,
where they signed the Virginia Nonimportation Resolutions, a

lengthy document of eight points. Summarily, it was to boycott all taxed material, including slaves, wines, or any other goods deemed appropriate. "Some Measures should be taken in their distressed Situation," it read, as a direct disobedience of the governor's power. George Washington, again, signed the pledge. So did Thomas Jefferson, Henry Taylor, Richard Henry Lee, and many others, all members of the House.[26]

Even before the resolution, many in the colonies were thinking beyond laws . . . and even beyond the sovereignty of the king. It was clear to many that these past years were nothing but proof of the difference between the Mother Country and colonies. So it was time to fight back.

Washington himself thought so, with an added caution. To wit, in a letter to wealthy planter, friend, and neighbor George Mason in early April 1769:

> *At a time when our lordly Masters in Great Britain will be satisfied with nothing less than the deprivation of American freedom, it seems highly necessary that something shou'd be done to avert the stroke and maintain the liberty which we have derived from our Ancestors; but the manner of doing it to answer the purpose effectually is the point in question.*
>
> *That no man shou'd scruple, or hesitate a moment to use a[r] ms in defence [sic] of so valuable a blessing, on which all the good and evil of life depends; is clearly my opinion; Yet A[r]ms I wou'd beg leave to add, should be the last resource.[27]*

This was a tougher stance than previously seen by Washington, much more explicit. The tension was becoming more than just an annoyance or an outrage. It was becoming a call to arms. But not yet. He was going through an evolution of thought—as were many colonists by 1769.

TENSIONS FLARED ON MONDAY, MARCH 5, 1770, A MONTH BEFORE THE REPEAL of the Townshend Act, when a rabble of Boston colonists and a misunderstanding of orders under British captain Thomas Preston led to the death of five and the injury of more. British soldiers were poorly, if at all, paid, and had to cast about for jobs. Boston's unemployed resented them, and mocked the British in their red uniforms, "Hey, bloodback! Hey, lobster for sale!" Colonists, young and old, mostly poor, pelleted snowballs and taunted the 29th British Regiment to fire on them. Words and insults were exchanged, someone gave the order to fire, and three colonists were killed instantly; two more died of their wounds soon after. A melee ensued, and the scene was a bloody carnage. Paul Revere's engraving of the Boston Massacre soon after was a hit, depicting sneering and shaded redcoats firing upon the helpless, unarmed civilians, as they begged for their lives, contorted, blood flowing onto the streets. It was the first major piece of propaganda of the colonies that rebutted the Empire. The term "massacre" was immediately thrown about.

The *Virginian Gazette* reported on the massacre on April 5, a month later, from an account of an unknown person who was there. The story took three pages, meaning most of that week's paper. A week earlier, the *Gazette* had noted a rumor spreading in Williamsburg about the attack, that "a fray happened lately at Boston, between some of the inhabitants and some of the soldiers, and that the latter fired upon, and killed several of the former. . . . We hope there is no truth in this report, and if there is, a few days will clear it up."[28]

A few days later, it was cleared up. The author of the April 5 article counted the dead, describing them. Samuel Gray, killed immediately, with "the ball entering his head and beating off a large portion of his skull." There was a black man, a dock worker named Crispus Attucks, also killed, with "two balls entering his breast, one of them [in] the right lobe of his lungs." There was James Caldwell, a "mate of Capt. Morton's vessel," shot in the back. There was Samuel Maverick, a seventeen-year-old, shot through the stomach, who died in agony the morning after. Patrick Carr, an Irish immigrant, died two weeks after a ball went through his hip. Others were critically injured: Christopher Monk and John Clark, both about seventeen years old, who were shot above the kidney and above the groin, respectively; John Green, a tailor, shot under the hip; Robert Patterson, shot in the arm; David Parker, a young boy apprentice to a wheelwright, shot in the thigh.[29] Despite the relatively impassive tone of the paper—it tried to keep to the facts—it still painted a scene of carnage, blood in the snow of Boston. Later, attorney John Adams successfully defended the British soldiers in court, attacking the rioting mob as "saucy boys."

THESE YEARS OF PARLIAMENTARY ACTS, TAKEN TOGETHER, SPARKED AN UN-derlying, seething anger among those in the colonies. Perhaps it had always been there—and the 1760s just uncovered it. Perhaps the inability of Parliament to listen to the colonies exacerbated it; on the other hand, perhaps it was the colonies' growing refusal to help debt-heavy Britain.

But there was a distinct identity emerging, one separating from the Crown and from Europe. An identity that was distinctly American. And Mary and George Washington were in the heart of it, in the heart of Virginia, where soon, war would come.

AFTER THE REPEAL OF MOST OF THE TOWNSHEND ACT, THE COLONIES FELL into relative calm. A new British prime minister, Frederick North, saw fit to ease some tension. But just some. People continued to go about their days, gossiping as before. They continued to view news and events of England with interest, and any curiosity with a sense of importance.

One such curiosity was Maria Theresa, of Chester, England, a twenty-seven-year-old woman who was described as "the amazing Corsican Fairy, who has had the honour of being shewen three times before their Majesties." The *Virginia Gazette* wrote about the woman, calling her the "most astonishing part of the human species . . . She is only 34 inches high, and weighs but 26lbs. A child of two years of age has larger hands and feet."[30]

With all the tension and cultural upheaval brewing, people in the colonies still were consumed with curiosity. Not even a year had gone by since the Boston Massacre, and news of an unusually small woman in England took more space than some advertisements for land.

SUCH WAS THE WORLD OF MARY WASHINGTON AT THE FREDERICKSBURG house. The interest and all-encompassing anger affected many colonists, from plantation owners to lawyers to merchants to townsfolk. One could not avoid hearing news of the Boston Massacre or the effects of the Stamp Act.

Or, perhaps, she heard the news, and simply didn't care. Taxes were a way of life, and the underlying philosophy of *independence* and *sovereignty* did not register as important to her. The Crown was the Crown, Parliament was Parliament; it didn't matter if it was across the ocean or not. Mary had a reputation of being anti-autonomous to her own adult children; perhaps she thought autonomy of the colonies was not something with which she concerned herself.

So says biographer Virginia Carmichael: "She had been taught since childhood to look upon England as good and great. England was the Mother Country who wanted prosperity for her colonies. Still, here was George, Fielding, her other sons, Hugh Mercer, and even her beloved Dr. Mortimer, the family physician, opposing England's policies."[31]

HER RELATIONSHIP WITH HER ELDEST SON, DESPITE THE TURMOIL IN THE colonies and his increasingly hardened political views, did not wane. He stayed with his mother at her newly moved-in house in April 1772, though he did not dine with her, and he gave her 8 pounds to assist his brother Charles.[32]

Again, in September 1772, George stayed in Fredericksburg with his mother for three days, meeting Charles, his sister and brother-in-law Betty and Fielding Lewis, and fellow veterans of the French and Indian War. On the last day, he gave Mary

30 pounds "in the presence of my Br. Charles" for some unspecified reason.[33] If she had been complaining about the move, perhaps it was a little cash to tide her over. In November, he again gave her 15 pounds, though he did not stay at her place, bedding instead at the home of Colonel Henry Lee.[34]

Every time he visited Fredericksburg, he made a conscious effort to visit his mother.

ACCORDING TO THE TEA ACT OF MARCH 1773, COLONIAL TEA COULD BE PURchased only from the British East India Company. The company was financially strapped but had warehouses full of tea in London. Parliament imposed the act to help them out of their monetary difficulties. This hit a nerve with the already agitated Americans. The tea tax of the Townshend Act was still in force, and the usual means of smuggling via the Dutch was no longer viable with the tightening British trade. Within the year, on December 16, Boston residents, dressed as Native Americans, boarded three East India Company ships under the cover of darkness, and dumped their entire cargo, up to 342 chests. It was as much a protest against an injustice as a practical means stopping the shipment of cargo.[35]

The Boston Tea Party was yet another spark to light the shortening fuse. "One universal spirit of opposition animated the colonists from New Hampshire to Georgia," wrote John Corry.[36] Colonists from Samuel Adams of Massachusetts to local unnamed Virginians defended the action. A decade of tension between the two governments had passed, and one *Gazette* article placed the blame for the Tea Party on former governor of Massachusetts Thomas Hutchinson, an incompetent and fierce Crown loyalist, even by colonial standards.[37]

One person who questioned the actions of the unidentified Bostonians was George Washington himself. He had known the captain of one of the ships, Captain James Bruce, who had dined with him years earlier.[38] He saw destruction of property as hurtful to the cause of sovereignty; being rabble-rousers and destroying property that was not theirs legally could only inflame the British.[39]

Inflame them it certainly did. They were furious, passing the Boston Port Bill, as part of the larger Intolerable Acts, on March 25, 1774, which closed the port and stopped shipment of all cargo, "goods, wares and merchandizers, at the town of Boston, or within the harbour."[40] It was no less than martial law, an invasion to suffocate Boston for their insubordination. A bitter and uneasy quietude settled over Boston.

ON MAY 27, VIRGINIA FOUGHT BACK. THE STILL DISSOLVED HOUSE OF BURgesses met in Williamsburg, again at Raleigh Tavern, and at ten in the morning unanimously agreed that "in Support of the constitutional Liberties of AMERICA, against the late oppressive Act of the British Parliament respecting the Town of Boston, which, in the End, must affect all the other Colonies . . . [tea] ought not to be used by any Person who wishes well to the constitutional Rights and Liberty of British America."

In other words, the legislative body agreed not to drink tea. It was drawing a red line in the sand, one that many thought the British Empire had crossed long ago.

The House similarly called for all to pray and fast on June 1, when the bill would take effect, to be a day "of Fasting, Humiliation, and Prayer, devoutly to implore the Divine Interposition for averting the heavy Calamity which threatens Destruction to our civil Rights, and the Evils of civil War. . . ."

The call for prayer continued, asking that God "give us one Heart, and one Mind, firmly to oppose, by all just and proper Means, every Injury to American Rights."[41]

It worked. George Washington, on June 1, a day observed to be clear and hot much like the colonists' temper, noted simply in his diary that he "went to Church & fasted all day."[42]

The city of Fredericksburg followed suit, boycotting tea and the company's products in response to the Boston Port Bill. One resident, John Harrower, writing to his wife, said, "As for Tea there is none drunk by any of this Government since 1st June last, nor will they buy a 2ds worth of any kind of east India goods, which is owing to the difference at present betwixt a tax the Parliament of great Britton and the North Americans."[43]

So what did Mary do, in the wake of this political upheaval and boycott?

She bought tea, paying a certain Robert Broom 18 shillings for Hyson tea.[44] "Tea—hot, with cream and sugar, and served in delicate cups—was one of life's actual necessities in the mind of elderly ladies whose mothers in England had been born drinking it," wrote Nancy Turner.[45] Many colonists saw the tax on tea as outrageous for precisely that reason—it wasn't a delicacy or some pleasure, but akin to water. Though the transaction occurred before the boycott—May 18—it was a deliberate move to express indifference on her part.

To George, though, this was "the cause of America," a rallying cry to unite the colonies under one banner. This was a brutal attempt "to fix the Shackles of Slavry [*sic*] upon us."[46] There was no more diplomacy, no more subtlety. It was pure injustice. For a colonist to go without tea was unheard of.

But it was time for something more, something more than, as Washington put it, "cry[ing] for relief, when we have already tried it in vain."[47] Whatever the colonies had done was not enough. It was time for action, not letters to the king.

VIRGINIA WAS NO STRANGER TO HEAVY RAINFALL AND HIGH WINDS, BUT perhaps that summer of 1774, when a northeast hurricane hit land, this was seen less as a natural event and more as a foreboding omen. In late August, said one man from Nomini Hall in Westmoreland County, "A violent Gust of Wind, Rain, & some Thunder we had about twelve o Clock, the Country seems to be afloat. . . . This is a true August Northeaster."[48]

With the thunder came Mary's long-held fear of storms.

Soon, it wouldn't only be the thunder in the clouds she'd hear, but the thunder of something else, weapons that would strike fear into men and women, husbands and wives, and sons and daughters. This thunder wouldn't be the wrath of the Almighty but the wrath of man. And for Mary, this thunder, from the explosion of revolution, would be as terrifying as the lightning from above.

After the storm, George would leave the safety of Mount Vernon and the comfort of his wife and family for the city of Philadelphia, over 130 miles away. With him was his slave, Billy Lee, and fellow members of the Virginia House, firebrand Patrick Henry and lawyer Edmund Pendleton.

ALSO IN VIRGINIA IN THE SUMMER OF 1774, THE CITIZENRY OF ESSEX County—just down the road from Mary Ball Washington's Fredericksburg—passed the Colonial Resolutions "to consider the present dangers which threaten ruin to American liberty." In details, these Virginians issued a seventeen-point declaration against the taxation from Parliament, against the consumption of East India tea, against exports to England, against imports from England, and funds were to be raised for the "poor of Boston," which, with the British blockade, was suffering a depression. However, the respect for King George III was also articulated. Respected judge and member of the House of Burgesses Muscoe Garnett of Ben Lomond was one of several local citizens entrusted with raising the funds and delivering them to Boston.[49] The Resolutions, impressively written, became more evidence of the growing distrust across the colonies—it was not just a simmering revolt by Boston. Dismay was spreading across the land.

IT WAS NOT UP TO GEORGE TO MAKE WAR WITH THE MIGHTY BRITISH EMpire, whether he wanted to or not. Frustrations toward the indifferent Parliament had reached a crescendo. Therefore, a meeting of the colonies had to be called, and on September 5, 1774, at Carpenters' Hall in Philadelphia, they met. At a very short distance from the edge of the Delaware River, this newly built two-story hall was the ideal spot for such a momentous meeting. This was more than just a meeting of Virginians, or a meeting of locals: fifty-six delegates from all the colonies, save for Georgia. It was the first of a series of Continental Congresses, held secretly and behind closed doors. "Congress," in colonial America, was simply a meeting or gathering; this was not a governmental body, like a parliament.

A month earlier, in August, over one hundred delegates met at the Virginia Convention in Williamsburg. They voted for seven delegates to represent the state in Philadelphia. Washington, now forty-two years old, was one of those appointed on Friday, August 5. Thomas Jefferson noted that a "shock of electricity" was felt at the convention.[50] The men knew what they were doing, knew the severity of it. "We never before had so full a Meeting of Delegates at any one Time, as upon the present Occasion," Washington wrote to Jefferson that same day.[51]

It was momentous.

But it was just an inkling of what was to come.

On the ninth of August, heading home to Mount Vernon, Washington went to Fredericksburg, where he stayed with Fielding and Betty Lewis for one night. While there, he gave his mother 20 pounds.[52]

Maybe, on this rainy day, he told Mary that he had been appointed a delegate to the Continental Congress. Political events did not interest her. The trip to Philadelphia was arduous, but he was not going to battle, so she would not be concerned for his safety. She was seemingly indifferent to his interest in the fate of the colonies. When her son stopped by and told her what had transpired, how did she react? No one knows for sure, but perhaps that old motherly protection kicked in. Perhaps she did not want to see her son, once again, leave her for some political rabble.

But with either her blessing or an indifferent sigh, he nevertheless went.

FROM MASSACHUSETTS THERE CAME SUCH HEAVYWEIGHTS AS COUSINS JOHN and Samuel Adams, Thomas Cushing, and Robert Paine. "It was their people who had most provoked Parliament to be high-handed and aggressive," wrote Woodrow Wilson, many years later. "All the continent and all England had seen how stubborn was the temper, how incorrigible the spirit of resistance, in that old seat of the Puritan power, always hard set and proud in its self-willed resolution to be independent."[53]

Certainly, the fervor that was raging through all colonies originated farther north, in New England, with Boston taking a brunt of the punishment. But the Continental Congress was not to lay blame or to exclude or damn Adams and company, or chastise them, or throw them to the Crown and beg for mercy on all others. They were to embrace them. "The Distinctions between Virginians, Pennsylvanians, New Yorkers, and New Englanders, are no more," declared Patrick Henry.[54]

It was not an easy path or an easy two-month convention. There were arguments, debates, and distrust. The interests and needs among the colonies were different; some delegates chose more radical action.

On one end, the more moderate, while clearly against the oppressive laws of Parliament, nonetheless wanted to keep the colonies distinctly British. There was no "natural right" to freedom, some even argued. Said John Rutledge: "Our Claims I think are well founded on the british [*sic*] Constitution, and not on the Law of Nature."[55]

On the other side of the spectrum were firebrands such as Patrick Henry and Richard Henry Lee, fellow Virginians. Declared Patrick Henry, "I am not a Virginian but an American. . . . Government is at an End."[56]

Ultimately, on October 14, 1774, the group decided to issue a series of declarations and resolves, saying that "the inhabitants of the English Colonies in North America, by immutable laws of nature, the principles of the English Constitution . . . are entitled to life, liberty, and property, and they have never ceded to any sovereign power whatever a right to dispose of either without their consent." This included their ancestors and it included immigrants, not just those native born, and it included their descendants. The colonists had the right to assemble, to air grievances, to petition the king.[57] It was not a declaration of independence by any means, but it was a couple of steps toward it. British goods were to be boycotted, and, if situations were not settled by next year, the congress would again meet.

Washington, during all this, acted "like a mature candidate," according to Ron Chernow. He attended churches of all denominations—from Catholic to Quaker—meeting and befriending all.[58] He did not want independence like his friend Patrick Henry, saying flat out that "no such thing is desired by any thinking man in North America."[59] He even gambled, winning 7 pounds.[60] While there, he did not forget his filial duties, buying specifically for his mother a cloak for over 10 pounds and a riding chair for 40.[61] He would not forget that.

The First Continental Congress adjourned on October 26, 1774; by the end of the month, Washington returned home to Mount Vernon.

IN FREDERICKSBURG THROUGHOUT THE MONTHS OF AUGUST, SEPTEMBER, and October, Mary would have gone through her usual motions for that period of time in the year. And as the fire of liberty was brewing in Philadelphia, in Fredericksburg "the great Race" took place.[62] For four days, starting precisely at eleven in the morning from October 4, a Tuesday, to October 7, a Friday, horses and mares of various ages and sizes raced on four-mile tracks.[63] On the fourth, William Fitzhugh's horse Regulus won; Honorable John Tayloe's mare Single Peeper won on the fifth. On the sixth, a Thursday, Fitzhugh won again, with his horse Kitty Fisher; and finally, on the seventh, Fitzhugh won yet a third time, with his horse named Chestnut. The news took half a column in the paper.[64] It was an immensely popular event to commemorate the seasons, with large prizes of pounds or guineas distributed.

Life went on in Fredericksburg. People came and went. Breweries went on sale.[65] Resident Jacob Whilly listed for rent a tavern, stable, kitchen, and bake-house for up to a year.[66] In late November, William Porter made quick work of selling garden seeds for the next spring.[67]

And so on. Mary likely saw many of these events firsthand, or read or heard about them.

Winter came and went. George noted December as a particularly heavy one; snow hit the thirteenth, twenty-third, twenty-seventh, and twenty-eighth of the month, with temperatures ranging from "not very cold" to "very cold." January, on the other hand, was relatively calm, with only a mix of rain and snow on the nineteenth, but only a handful of cold or cool days themselves, and February likewise had a couple of snow days, but was, overall, pleasant.[68] George himself, at least on the surface, seemed to be relatively calm before the proverbial storm.

This year 1775 would bring more changes, more drastic, more radical.

IT WAS MARCH 23, 1775. IN RICHMOND, VIRGINIA, AT HENRICO PARISH Church, an Anglican church built thirty-five years earlier. The weather was "Cloudy & Chilly—with appearances of Snow—wind being Easterly but none fell. Afternn. clear," according to Washington. In his diary, he wrote simply that he "Dined at Mr. Patrick Cootes & lodgd where I had done the Night before."[69]

For three days, the Second Virginia Convention had met here, hearing and ratifying the Continental Congress's resolutions. It was Patrick Henry, that same tough-minded fellow Virginian, who took the floor. Voice echoing and bellowing, all eyes on him, there was no question what he wanted for the colonies. "We must fight!" he yelled. "Is life so dear, or peace so sweet, as to be purchased at the price of chains and slavery? Forbid it, Almighty God! I know not what course others may take; but as for me, give me liberty or give me death!"[70]

THREE MONTHS EARLIER, GEORGE WASHINGTON, GEORGE MASON, MARTIN Cockburn, and several others proposed a plan to raise ammunition and funds for a militia to protect the rights of colonists.

Now, Henry's fiery speech put that to use. The assembly of Virginian delegates voted and agreed that a militia must be formed:

Virginia must be placed "into a posture of defence," and a petition for a group of volunteer militiamen was called. A tax was to be placed for the gathering of war supplies, and the delegates were to "prepare a plan for the encouragement of arts and manufactures in this Colony"—meaning, prepare for a war.[71]

THERE WERE OTHER PRIORITIES, HOWEVER, OTHER DUTIES TO BE DONE. Maternal priorities. In late April, George totaled all the money he had lent to or spent for Mary since 1771. The last entry, from March 30, was her yearly allowance of 30 pounds. In total, he had spent 434 pounds, 11 shillings, and 8 pence for her, not an unsubstantial amount. This did not include the various small loans, uncounted. This was simply allowances and rent of her land. Gifts and such could add up.[72]

YEARS LATER, WASHINGTON SAID THAT GREAT BRITAIN "UNDERSTOOD HERself perfectly well in this dispute, but did not comprehend America."[73] In another letter five years later, this time to Joseph Jones, he said that the Mother Country "thought it was only to hold up the rod, and all would be hush!"[74]

England underestimated the colonies' call for equality, and underestimated the collective frustration that had built up over decades. It wasn't just a town here or a rogue colony there, it was the collective thirteen. The colonies were not going to tolerate the parliamentarian or royal abuse anymore. They were not going to accept the breaking of laws and traditions. They were going to fight for a voice, for representation, for sovereignty.

Independence was another matter. That fight would not begin . . . yet.

Chapter 10

Off to War

THE AMERICAN REVOLUTION

1775–1783

---

*"I shall be ready to receive the misled
with tenderness and mercy."*

George Washington returned to Mount Vernon from Fredericksburg on the last day of March 1775, after dropping off his yearly allowance to Mary and dining with Fielding and Betty Lewis. On April 19, 1775, the weather was clear, but a hard wind from the west battered George's Mount Vernon home. That night, two of his indentured servants, twenty-year-old Thomas Spears of Bristol and thirty-year-old William Webster of Scotland, sailed away from the plantation, taking a small yawl with them. George offered a hefty reward for their return.[1] Indentured servants and slaves were prone to escape, to ironically seek that same freedom that the colonists and landowners wanted (in fact, Webster himself had previously run away); but overall, April 19, 1775, was normal.

In Massachusetts that day, something entirely different happened. There was no preparation for war. There was war. In the early dawn hours, at 5:00 a.m., British troops led by Commander Francis Smith, at least four hundred in number, entered a town northwest from Boston named Lexington with the intent of arresting and confiscating the revolutionary leaders and armaments. A battle broke out, with the band of misfit militiamen surprisingly driving out the professional redcoats. About eighteen Americans were killed or wounded. Two hours later, in Concord, farther west, the same British companies entered the town and were similarly repelled by what were becoming known as the Minute Men. That day, there were one hundred colonist causalities, but over two hundred British were dead and wounded. It was a decisive victory.[2]

Washington himself was "sobered and dismayed" by the news; writing to George Fairfax in late May, he noted that "unhappy it is though to reflect, that a Brother's Sword has been sheathed in a Brother's breast, and that, the once happy and peaceful plains of America are either to be drenched with Blood, or Inhabited by Slaves. Sad alternative!"[3] Whatever peace they had built up or tried to build, whatever pleas to the king, it was for naught.

War had come to America. In lore, the firing by the Americans at Lexington and Concord became forever known as "the shot heard 'round the world."

THE DATE OF THE SECOND CONTINENTAL CONGRESS WAS ESTABLISHED AT the close of their first meeting and could not have come at a better time. Delegates met at the Pennsylvania State House in Philadelphia on May 10, with a simple question: since no tension had abated, what should they do? The answer, however, was not so simple. Unlike the first, this congress would last until March of 1781, a period of nearly six years, through war, through upheaval, and through independence.

Militiamen and civilians welcomed incoming delegates at the outskirts of Philadelphia, making this a different and more popular—perhaps more urgent—congress. "The spirit of almost everything that day seemed encouragingly different" than those eight months earlier.[4]

For its day, the congress worked quickly.

On Wednesday, June 14, 1775, a Continental Army was formed.

The next day, a mere month after first convening, George Washington, dressed in his red military uniform, was unanimously appointed commander in chief and general of all the combined Continental forces, first nominated by Thomas Johnson, delegate of Maryland. Other nominees included John Hancock, president of the congress, who John Adams said had "an ambition" to lead.[5] Hancock was crestfallen.

It was not an easy vote, or a vote by acclamation that everyone wanted to support. The congress was worried. As honorable and valiant as Washington was, to appoint a commander of the forces would be yet another no-turning-back move for the colonists. Further, and perhaps more important, was a Virginia planter such a perfect choice to head a Massachusetts-led rebellion?

"What shall We do?" asked Samuel Adams of his cousin, in private; to which, John replied, "I am determined this Morning to make a direct Motion that Congress should adopt the Army before Boston and appoint Colonel Washington Commander of it."[6]

When the congress met again, John Adams did exactly that. As he wrote in his diary, "I had no hesitation to declare that I had but one Gentleman in my Mind for that important command, and that was a Gentleman from Virginia who was among Us and very well known to all of Us, a Gentleman whose Skill and Experience as an Officer, whose independent fortune, great Talents and excellent universal Character, would command the Approbation of all America, and unite the cordial Exertions of all the Colonies better than any other Person in the Union."

Hearing his name, George Washington fled from the room without saying a word.

Several delegates immediately raised objections, according to Adams, and Hancock himself fell into despair, feeling "mortification and resentment" at not being named.

With support so much greater than opposition, those who opposed were persuaded enough to vote aye.[7] Any objection to a Virginian leading, if any did exist, was squelched. A wealthy colony perfectly situated in between the north and south, the oldest of the colonies, to boot, with mighty ports and resources, was as good if not better an ally as any.

Washington's diary noted a simple entry, no different in tone than any other: "Dined at Burns's in the Field. Spent the Eveng. on a Committee."[8] Not a whisper about being appointed to take on any impossible task: to lead a ragtag group of volunteers against the most powerful army and navy in the world. But his mother was also known for keeping things close to her vest, as well.

Mary Ball Washington's son had been chosen to lead a revolt against the British Empire.

ON THE SIXTEENTH OF JUNE, WASHINGTON, WITH GREAT RELUCTANCE AND honesty of fear, accepted his nomination:

"Tho' I am truly sensible of the High Honour done me in this Appointment," he said to the assembled men, "yet I feel great distress, from a consciousness that my abilities and military experience may not be equal to the extensive and important Trust. However, as the Congress desire it, I will enter upon the momentous duty, and exert every power I possess in their service, and for support of the glorious cause. . . .

"But, lest some unlucky event should happen, unfavorable to my reputation, I beg it may be remembered, by every Gentleman in the room, that I, this day, declare with the utmost sincerity, I do not think myself equal to the Command I am honored with."[9]

One story, told secondhand, had Washington, upon accepting, walk up to Patrick Henry, fellow Virginian and not-so-fellow radical. Tears filled his eyes as the responsibility hit him. Tears of fear, nervousness, duty, honor. "Remember, Mr. Henry, what I now tell you," George said, quivering. "From the day I enter upon the command of the American armies, I date my fall, and the ruin of my reputation."[10]

On June 19, his commission as "General and Commander in Chief of the army of the United Colonies" was officially signed by Hancock. George Washington, forty-three years old, was officially commander of the entire thirteen colonies. The first order of business was to refuse a salary, accepting expenses only.

LETTERS FLEW. TO HIS BELOVED MARTHA, HE WROTE ON JUNE 18, "I AM NOW set down to write to you on a subject which fills me with inexpressable concern—and this concern is greatly aggravated and Increased when I reflect on the uneasiness I know it will give you." He continued, "It was utterly out of my power to refuse this appointment without exposing my Character to such censures as would have reflected dishonour upon myself, and given pain to my friends."[11] It is an eerily similar theme from his letter to Mary twenty years earlier, when appointed commander of the Virginia Regiment during the French and Indian War. It was humbling, certainly, but this was a cause greater than himself or any one man, and to refuse outright due to fear would be rebutting everything for which they'd worked.

To his brother-in-law, the husband of Martha's sister, Burwell Basset, he wrote, "I am now Imbarkd on a tempestuous Ocean from whence, perhaps, no friendly harbour is to be found. . . . [This position] is an honour I by no means aspired to—It is an honour I wished to avoid."[12]

To his adopted son John Parke Custis, he wrote similarly. As the oldest of Martha's surviving children, Custis had an additional responsibility. "My great concern upon this occasion, is the thoughts of leaving your Mother under the uneasiness which I know this affair will throw her into," Washington confessed. "I therefore hope, expect, & indeed have no doubt, of your using every means in your power to keep up her Spirits, by doing every thing in your power, to promote her quiet."[13]

To his brother John Augustine in Westmoreland County, he penned his opening, "I am now to bid adieu to you, & to every kind of domestick ease, for a while."[14]

To Martha his love, again, on June 23, the day he departed Philadelphia for Boston, he wrote a quick letter:[15]

Phila. June 23d 1775.

My dearest,

As I am within a few Minutes of leaving this City, I could not think of departing from it without dropping you a line; especially as I do not know whether it may be in my power to write again till I get to the Camp at Boston—I go fully trusting in that Providence, which has been more bountiful to me than I deserve, & in full confidence of a happy meeting with you sometime in the Fall—I have not time to add more, as I am surrounded with Company to take leave of me—I retain an unalterable affection for you, which neither time or distance can change, my best love to Jack & Nelly, & regard for the rest of the Family concludes me with the utmost truth & sincerety Yr entire

Go: Washington

He left Philadelphia that morning, knowing completely the task that had been handed him and what would happen should he fail.

To all, he sent similar notes: his reluctance in accepting; the ache it would bring to his family; that the congress had adopted a Continental currency to help raise payment for troops and materiel; the appointments of Major Generals Charles Lee, Artemas Ward, Philip Schuyler, and Israel Putnam. Each of these letters, while similar in words and themes, could not be read with a sense of calm and serenity. There was a real sense of urgency.

The next years would be frenetic, and no fact highlighted this more than the absence of any diary entries by George Washington, so copiously penned for decades prior, from June 19, 1775, to early 1780.

The coming days irrevocably changed his routine.

ONE FAMILY MEMBER WHO WAS CONSPICUOUSLY ABSENT FROM THE LIST OF letters he had written was his mother, the same woman who worried about his trip to Barbados, his trip to Ohio, his trip to sea.

The same woman who asked him for money and allowances. The same woman who couldn't help but worry. If a letter did exist, it is long lost; whether through the march of time, or the pillaging of souvenir seekers and Civil War soldiers, or Mary discarding the letter, none has been found.

What did she think? When did she find out? Most likely, she heard it from Betty. The *Virginia Gazette* first reported the news on July 6, with one sentence, "The honorable congress have appointed GEORGE WASHINGTON, esquire, of Virginia, generalissimo of the American army." In the same issue was news of his leaving Philadelphia for Massachusetts, where he would personally command the stationed army.[16] The *Gazette* was four days late in publishing, as Washington had arrived at Cambridge, Massachusetts, on July 2. Whatever news Virginia was going to hear would be delayed.

Mary had experienced this same fear before, so many times. But this time it wasn't just her eldest that was off to war. So too was her grandson, George Lewis, Betty and Fielding's son. So too was Hugh Mercer, her neighbor and onetime doctor. So too was Bushrod Washington, son of John Augustine, and George Steptoe Washington, son of Samuel Washington; another of Betty's sons, Robert Lewis; and her husband, Fielding, who used his five-plus ships for patrolling the Rappahannock, as well as smuggling goods. Others close to Mary were off to war as well: John Paul Jones and George Weedon, both Fredericksburg neighbors. While Mary's youngest son, Charles Washington, never saw battle, he was still an active participant, taking a politician's role in Fredericksburg. Similarly, while her next-door neighbor, James Mercer (no relation to Hugh), never saw battle in the Revolution, he did not shy away from participating behind the scenes.

The city of Fredericksburg itself was in a revolutionary fervor. The classic exhortation "God save the king" was replaced, proactively, with "God save liberties of America," even at official city committee meetings.[17] Fredericksburg had already been chosen in July 1775 as the site of a gunnery to manufacture needed war materiel; Fielding Lewis put his all into its upkeep. Mary would have been caught up in this fervor, in one way or another. In such a time, it would have been impossible to plead ignorance or bury your head in the sand and pretend the world you knew wasn't ending.

A curious chronology error existed at this point in time. George Washington Parke Custis was perhaps the source of it. He noted that his grandfather, "previously to his joining the forces at Cambridge [in early July of 1775], removed his mother from her country residence to the village of Fredericksburg, a situation remote from danger, and contiguous to her friends and relatives."[18] Author Mary Terhune, in the late nineteenth century, said similarly; after the Battle of Bunker Hill in June of that year (1775), George "begged" his mother to move closer to family.[19] While Fredericksburg offered some safe haven from the nearby Rappahannock and the farther Potomac, it was still a major trade city, square in the middle between north and south Virginia. It wasn't such a dramatic move as Custis let on, either. It was only a mile's distance between Home Farm and her house. Across a river, yes, but a small one at that juncture. However, and most obviously, the threat of war did not result in her move; she had moved there three years earlier. There were rumbles, but no one was expecting a revolution by 1771.

But Custis and Terhune, prone in this case more to style and themes than facts, demonstrated here their belief that the general deeply worried about the elder Mary, so much so to upend her comfort for her safety. And that was perfectly true: she was elderly, she had suffered a bad case of the flu the winter before, and she was alone, caring for a by now failing farm.

NO GOODBYE WAS SAID; THERE WAS NO TRIP DOWN TO FREDERICKSBURG BE-fore setting off. Mary, mother of Washington and nearly seventy years old, could not offer a farewell or grant a blessing. She instead had to wait to hear word of her eldest son "for a period, the duration and events of which no mortal vision could even faintly discern," wrote historian Margaret Conkling in 1850.[20] If Mary had not cared about the tensions before, she certainly did now. Not only was her son to lead a war, but it was a war against the very same people who had led him. It was one thing to fight the French or the Indians or the non-English; it was another thing altogether to fight fellow Englishmen.

When she heard of his command and his going off to fight yet another war, it was possible, said Michelle Hamilton, manager of the Mary Washington House in Fredericksburg, that there "would have been a heavy sigh, as it's not the path she wanted her son to go on. Being in the military is risky," and she recognized that. But it was out of her control, which would have only been more nerve-racking. "There would've been a heavy dose of concern." After all, she would have heard of the letters from her son in the French and Indian War, when he wrote how much he enjoyed the sounds of bullets ricocheting and flying past his head. Concerning, especially for a mother.[21]

THE REVOLUTION RAGED UNOFFICIALLY IN THE REMAINING MONTHS OF 1775. By December, a dozen or more battles and untold skirmishes had taken place in the northern colonies. From what was a single fight in April outside of Boston to a war, there were now sieges and bloody continuous fighting. Many of these battles were victorious for the Patriots; some, like the fabled Battle of Bunker Hill in Charlestown, Massachusetts, were not, with a strategic and decisively early British victory. Though about 1,050 British troops were killed or wounded, a devastating number, about 450 American casualties were reported too, with the British successfully capturing the nearby Charlestown Peninsula.[22] George Washington was not part of the battle, but at the time in command at New York City.

THE CONTINENTAL CONGRESS IN PHILADELPHIA HAD TRIED AND FAILED TO wage a final peace. In the so-called Olive Branch Petition, the congress agreed to a supplication to King George III with a tactful letter: *The colonies were unhappy with Parliament's and your ministers' policies, not with you or your own.* "Knowing to what violent resentments and incurable animosities, civil discords are apt to exasperate and inflame the contending parties, we think ourselves required by indispensable obligations to Almighty God, to your Majesty, to our fellow subjects, and to ourselves, immediately to use all the means in our power, not incompatible with our safety, for stopping the further effusion of blood, and for averting the impending calamities that threaten the British Empire." Delegates signed and agreed to the petition, asking that peace be made and that the king live long. From Virginia, signers included Patrick Henry, Richard Henry Lee, Edmund Pendleton, Benjamin Harrison, and Thomas Jefferson.[23]

It was a last-ditch effort for peace. As radical as some of the Massachusetts rebels were, many Virginians, including Patrick Henry, did not want all-out war. That would devastate the colonies, they believed.

Whatever the king thought when he read the petition, it failed. On Friday, October 27, 1775, in an address to both the House of Lords and the House of Commons in Parliament, King George III laid down the gauntlet. "It is now become the part of wisdom, and (in its effects) of clemency, to put a speedy end to these disorders by the most decisive exertions," he exclaimed. "For this purpose, I have increased my naval establishment, and greatly augmented my land forces; but in such a manner as may be the least burthensome to my kingdoms."

He continued, making it perfectly clear his intentions: "When the unhappy and deluded multitude, against whom this force will be directed, shall become sensible of their error, I shall be ready to receive the misled with tenderness and mercy."[24]

THE WAR WAS NOT ISOLATED TO ONLY THE NORTH DURING THESE MONTHS. It was a war for the thirteen colonies. Virginia would see significant first blood spilled in battle in November of 1775 at the Battle of Kemp's Landing (located in Virginia Beach). Kemp's Landing was the culmination of tension between Lord Dunmore, governor of Virginia, and the Virginia colonists. For several months, small clashes between the two forces resulted in wounded and dead. The British HMS *Hawk* at one point ran aground; ten seamen were captured and its captain killed.

Dunmore, fed up, landed over one hundred soldiers and volunteers near the Great Bridge in Virginia Beach on November 14; that, in turn, put the militia of Princess Anne County on alert for an ambush. The next day, November 15, the two clashed, 170 colonial militia versus 100 infantry and 20 Loyalists. It was a decisive British victory; though the number of troops was small and the number of casualties smaller (seven dead or wounded Virginians), it was significant in result: here, the loud Virginian colonists were overwhelmed by the mighty British infantry. One British grenadier was wounded when he hurt his knee.[25]

The governor, hearing of victory, issued a proclamation, written a week prior: the Colony of Virginia was in a state of rebellion. All loyal subjects were to put it down, to "resort to his Majesty's STANDARD, or be looked upon as traitors." Further, and more radically, he issued a promise not to the white colonists, but to the slaves and indentured servants: "I do hereby further declare all indentured servants, Negroes, or others . . . free, that are able and willing to bear arms, they joining his Majesty's troops, as soon as may be."[26] It was an attempt to create insurrection against the colonists.

It was his own Emancipation Proclamation.

The *Virginia Gazette*, which published the short order in full, gave lengthy commentary on it. It objected to the use of "rebel" as defenders, not attackers, of the Virginia Colony, and that such a term might deter others from joining the cause. How could one be a rebel if they were fighting *for*, not *against*, the welfare of Virginia? How could they be "rebels" when they had hardly fought at all? "We petitioned once and again, in the most dutiful manner, we hoped the righteousness of our cause would appear, that our complaints would be heard and attended to; we wished to avoid the horrors of civil war."

It continued, more emphatically: "We, my countrymen, are dutiful members of society; and the person who endeavours to rob us of our rights, *they are the rebels, rebels to their country, and to the rights of human nature.*" The italics were in the original.

To the proposal freeing slaves, it said they'd been "flattered with their freedom," but issued several warnings: first, what was to be done with those who weren't able to fight? The condition left off the elderly, the young, the women, the sick. If a father managed to escape, then the "masters [would] be provoked to severity." The proclamation, it said, "leaves by far the greater number at the mercy of an enraged and injured people." Slaves were property, and if the governor declared they no longer were, then what gave him that right? Certainly not the slaveholding colonists, whose law about runaway slaves or servants had been passed on from common law. Further, the *Gazette* believed that the British Empire, not the colonists, were the real extremists in slavery. At the end of the war, what then would happen to the slaves who fled to the Loyalist cause? "They will give up the offending Negroes to the rigour of the laws they have broken, or sell them in the West Indies, where every year they sell many thousands of their miserable brethren, to perish either by the inclemency of the weather, or the cruelty of barbarous masters. . . . Be not then, ye Negroes, tempted by this proclamation to ruin yourselves."[27]

Though minuscule compared to the number of militiamen or rebels, escaped slaves who heard Dunmore's promise enthusiastically joined for the Ethiopian Regiment, an easy task for Dunmore compared to the recruitment of other Loyalist regiments.[28]

AS IF A FERVOR WASN'T YET SWEEPING THE COLONIES COME JANUARY 1776, IT would only increase with the publication of an anonymous author, whose work, titled *Common Sense; Addressed to the Inhabitants of America*, became the most widely publicized and read work in American history. Its author, Thomas Paine, unknown at the time, "was an English radical, in his late thirties, who had failed at virtually every job he ever tried," as described by historian James Stokesbury.[29] Time was making more converts to the colonist cause, but so was reason, as Paine had written.

Common Sense was not subtle in its motive: independence from Great Britain, independence from the monarchy, establishing something new and different—a republican government. It was so radical that there were rumors half a year later both domestic and overseas that John Adams was the author.[30]

Within a month of its initial publication, three editions had been printed. Extracts immediately hit the papers. The *Virginia Gazette* ran a relatively short one, from the pamphlet's third section, "Thoughts on the Present State of American Affairs," where Paine wrote, "I challenge the warmest advocate for reconciliation to show a single [advantage] that this continent can reap by being connected with Great Britain. I repeat the challenge, not a single advantage is derived." He argued that any crop, or export, that they had would have been just as valuable to other nations without the bureaucratic mess of England. Further—and this resonated with those who had lived through the French and Indian War—any English war affected the colonies in trade and economy. Worst of all, war wasn't a one-and-done deal. There was no "war to end all wars," or real calls for permanent peace. Paine argued just the opposite, that the culture behind England was inherently warlike. "Europe is too thickly planted with kingdoms to be long at peace."[31]

A CURIOUS NOTE WAS MADE IN THE *VIRGINIA GAZETTE* ON JULY 12, 1776: IT was placed on the third page, in a single sentence: "The postmaster in Fredericksburg writes, as of last Wednesday, that, by a gentleman just arrived from Philadelphia, he had seen at an Evening Post of the 2d instant, which mentions that the Hon. the Continental Congress had that day declared the United Colonies free and independent states."[32] On July 2, the Continental Congress unanimously voted in favor of independence, a day that John Adams believed would be celebrated by subsequent American generations. Two days later, the Fourth of July, the inspired document was ratified. It opened,

> *When in the Course of human events, it becomes necessary for one people to dissolve the political bands which have connected them with another, and to assume among the powers of the earth, the separate and equal station to which the Laws of Nature and of Nature's God entitle them, a decent respect to the opinions of mankind requires that they should declare the causes which impel them to the separation.*
>
> *We hold these truths to be self-evident, that all men are created equal, that they are endowed by their Creator with certain unalienable Rights, that among these are Life, Liberty and the pursuit of Happiness.*

The document listed several dozen grievances, not against Parliament only, but against the king. It marked a turning point in American thought: all previous petitions, even a year earlier, believed the monarch to be the legitimate sovereign. It was Parliament and his ministers that were mucking things up.

No more diplomacy here.

The Declaration of Independence became a rallying cry, not just for the colonies but, as many believed, for all of humanity. Here, certain individual rights, created by God, trumped any human law.

THE REVOLUTIONARY WAR BECAME THE WAR OF AMERICAN INDEPENDENCE, making it less a fight against oppression and more a fight for freedom. A fight for America. As if the rupture was not apparent enough, on June 14, 1777, the Second Continental Congress voted to create a thirteen-striped, thirteen-stared flag, a unifying flag among all the colonies. The stars were to be "representing a new constellation," a new break from the Mother Country.[33] The Union Jack was forever furled.

The Declaration of Independence only divided and solidified party lines. The Revolution in general and the War of Independence in particular were fought by three camps: the Patriots, the British, and the Loyalists. All three groups had distinct advantages, disadvantages, ideologies, and backgrounds.

The Patriots—or rebels, if you asked any loyal subject—had an offensive war to fight, to fight *for* freedom, *for* independence, *for* a cause. The army consisted of former militiamen, most of whom were equivalent to normal infantrymen in Europe. There were hardly any grenadiers of note, and light infantry were introduced in May 1776. Cavalry was effectively nonexistent. The congress had allowed Washington to raise a fighting force upward of 27,500 men for one year, a sizable army, but the lack of cavalry and the British blockades made even the simplest task nearly impossible. What was wished for and what came to be were quite different, as by January of 1778, only 8,200 men were enlisted in the army.

It was an uphill battle.

The early Continental Army did not even have a simple uniform for its soldiers to wear. Originally it was the brown of the New England militia. In 1779, the infantry was ordered to be dressed in dark blue, with customizations by each state.

Good weapons, likewise, were hard to come by. Musket balls and powder that each infantryman had were subject to rain and bad weather, rendering them useless. Artillery and cannon were inferior as well in both quality and quantity. Inferior as it was, the cannon was instrumental in the struggle for independence. General Henry Knox was one of the few notable—and exceptional— men who handled the cannon with pride, at battles like Boston, Trenton, Princeton, and Monmouth.[34]

That uphill battle was looking more and more like a mountain. By the end of the war, it was estimated that about 150,000 men enlisted in the Continental Army, with an additional 145,000 in the militias, representing about 10 percent of the colonial population.[35] While that was sizable, it took more than numbers to win a war.

The British Army, experienced in foreign and overseas battles, had the advantage in nearly every respect. They had organization and efficiency. Their uniforms were tight and striking, the famous red looking like a tidal wave of blood across the valleys as men marched to battle. They had been called since the 1740s both "Bloody Backs" and "Lobster Backs," partly due to their uniform but also for their liberal use of the whip against criminal offenders. Men were coerced to join, and often soldiers themselves were criminals, offered freedom in return for their service. If that didn't work, then get them drunk and have them "agree" to enlist.[36]

The redcoats were formidable, and they had better weapons, including the sixty-year-old flintlock musket known as "Brown Bess," weighing fourteen pounds with a forty-two-or-so-inch barrel. It was accurate and deadly, to boot; one British colonel, George Hanger, wrote that it could "strike the figure of a man at 80 yards; it may be even at a hundred." Any more distance would sacrifice accuracy, and Hanger wrote that at two hundred meters, "You may as well fire at the moon."[37]

Foreign mercenaries assisted the British Army, only adding more power behind the already deadly force: about thirty thousand Germans fought for British control, the largest of which came from Hesse-Cassel.[38]

Finally, the Loyalists, that sort of mixed blend of American and British: American in culture, but British at heart. Their other label, the Tories, was perhaps uttered by rebels with disgust, the term going back a century. With all the demands that England and Parliament and the king had made on the colonists, they thought rebellion would foment. However, that certainly didn't mean a declaration of *independence* from the Crown was necessary.

Their beliefs were just as intellectual and just as philosophical as the Patriots', they thought. The divine right of kings had been a historical norm for centuries. God Himself anointed kings as rulers of nations, so to go against the rightful king would be to go against the Supreme Being.

Further, what would it really change? Liberty and individuality—if they were common among all peoples, as the rebels said, then should not they be common under a king? "You have taken some pains to prove," wrote "A. W. Farmer" (in reality, the bishop of Connecticut, Samuel Seabury), "what would readily have been granted you—that liberty is a very good thing, and slavery a very bad thing. But then I must think that liberty under a King, Lords and Commons is as good as liberty under a republican Congress: And that slavery under a republican Congress is as bad, at least, as slavery under a King, Lords and Commons: And upon the whole, that liberty under the supreme authority and protection of Great-Britain, is infinitely preferable to slavery under an American Congress." He continued, writing, "Man in a state of nature may be considered perfectly free from all restraints of law and government: and then the weak must submit to the strong. From such a state, I confess, I have a violent aversion."[39]

The Loyalists were always a part of the Revolution, just as much as the rebels. The first corps was formed in Massachusetts, in the fall of 1774, of about three hundred men. The actual fighters, a small percentage of the greater Loyalist cause, were almost unwelcomed in the war by both sides. Rebels, clearly, thought of them as traitors. The British, however, thought them ineffective, inferior to the professional British Army. The Tories' call to fight was delayed by Parliament, and when it was approved, the results were ultimately disappointing in both number and deed.[40]

SO WHERE DID ALL THE WOMEN GO?

Martha was left alone; Mary was left alone; so many wives and daughters and mothers and sisters were left alone, waiting to hear of their husbands' or fathers' or sons' or brothers' situation. If the men did not die from a musket or cannon, then they died of disease. In February 1777, camped at Morristown in New Jersey, Washington wrote that the smallpox that hit the soldiers—which he had been inoculated against due to exposure decades earlier in Barbados—was so great and sweeping, "we should have more to dread from it than from the Sword of the Enemy."[41] Further, smallpox was only one disease of many. Dysentery and generally poor health conditions led to the spread of infection and illness. The enemy was often the least of their concerns.

Though women did not take an active role in fighting, their role behind the scenes was just as vital. "Yes, Ladies. You have it in your power more than all your committees and Congresses, to strike the stroke, and make the hills and plains of America clap their hands," wrote William Tennent III of South Carolina.[42] One woman named Molly Pitcher was said to have participated in the Battle of Monmouth in June 1778, though her participation was believed to be more legend than truth.

From the start, they all realized sacrifices had to be made. Not just of their husbands, but of themselves as well. "Our ALL is at stake," wrote one Philadelphia woman to a Boston officer, soon after the bloodshed at Lexington and Concord, "and we are called upon by every tie that is dear and sacred, to exert the spirit that Heaven has given us in this righteous struggle for LIBERTY."[43] Man, woman, child, or elderly—all had to sacrifice. They were susceptible to the horrors of war, yet were still an important rallying cry for independence in all the colonies.

A simple look through documents of the day confirms this.

Before the war, back in 1774, the women of Pennsylvania were asked by their neighbor Virginians to boycott tea in support of the colonies.[44]

The second publisher to print the newly signed Declaration of Independence was a woman. Mary Katherine Goddard of Baltimore, Maryland, took an unusual role for that time period. She was the first female publisher in the colonies as well as the first female postmaster. She took the opportunity to print, with the permission of the congress, the Declaration with all typed signatories. At the bottom of the newspaper, at the place for the owner of the printing, she placed her own name.[45]

In September 1776, closer to home for Mary and Martha, the *Virginia Gazette* ran a short letter from Anne Terrel of Bedford

County "to the LADIES whose husbands are in the continental army." It urged their support for the war. Terrel wrote, "I am not only willing to bear the absence of my dear husband for a short time, but am almost ready to start up with sword in hand to fight by his side in so glorious a cause. But let us support ourselves under the absence of our husbands as well as we can." This included, she noted, making their own cloth and material, without the need for British trade.

"Let the tyrants of Great Britain see that the American Ladies have both ingenuity and industry," she wrote.[46]

Among the many ways to boycott and increase American-made manufacturing was through "spinning bees," to create fiber for clothes and other material. It was "almost an extension of their household work," but for a cause greater than their home or plantation, wrote one historian.[47] One pamphlet by Esther Reed of Philadelphia from June 1780, titled "The Sentiments of an American Woman," asked the colonial women to be like "the heroines of antiquity." Funds were also raised by organizations of women for the army, which were much needed.[48]

Despite Terrel's call for domestic support of their husbands, some women took more active parts in the war itself. Some had agency beyond staying at home. In April 1777 near Woodbridge, New Jersey, an unnamed woman, noticing a drunken Hessian soldier away from his regiment, went home, dressed as a man, and, with a gun, successfully took him prisoner and delivered him to a nearby Continental patrol.[49]

The women were on both sides of the fight, of course. In Philadelphia in 1779, a warning about "the wives of so many of the most notorious of the British emissaries" went out, suspecting them of espionage, "sending all the intelligence in their power, and receiving and propagating their poisonous, erroneous, wicked falsehoods here."[50] Three Virginian women that same year were captured near Suffolk, Virginia, and forced onto British ships, including "one of whom was on the point of being married."[51] It was unknown if they ever were returned; rape was common, a fear all too real for some, even those as young as ten years of age.[52] The British treated prisoners of war in the most terrible manner—women included.

THE MOOD OF THE COLONIES WAS TENSE. WILLIAMSBURG, VIRGINIA, LONG held to be remote from the booming population moving west, was no longer a viable capital city. With the James River to the south and the York River to the north, it was easily accessible by any invading British ship. At the urging of Thomas Jefferson, then governor, in 1779, the legislature agreed to move the capital away. After much bickering and debate about whether the new seat of state government should be Fredericksburg, Charlottesville, Staunton, or some other area, it was finally agreed in the spring of 1780 to move the capital to Richmond, founded some forty years earlier and some forty-five miles to the northwest, more inland and more secure—and closer to Mary's Fredericksburg.[53]

WAR CHANGED MEN. TO BE IN BATTLES AND LEAD OTHERS THROUGH SLEEP-less nights, military strategies to shrink the inevitable loss of life, to kill others who are trying to kill you. To see the carnage, the blood, the gore, the severed legs and arms, the bullet and cannon wounds that tore into men's chests and skulls, leaving nothing but pulp in their wake. To see plagues and hunger and sickness among people, not by the dozens but by the hundreds, if not thousands. That can change a person, physically, mentally, spiritually. It can age that person beyond the years that he lived.

So did it change George, the veteran?

One man, writing in the *London Chronicle* on July 22, 1780, wrote that Washington "is a tall, well-made man, rather large-boned, and has a genteel address. His features are manly and bold; his eyes of a bluish cast and very lively; his hair a deep brown; his face rather large, and marked with the smallpox; his complexion sunburned, and without much color. His countenance sensible, composed, and thoughtful. There is a remarkable air of dignity about him, with a striking degree of gracefulness. He has an excellent understanding, without much quickness; is strictly just, vigilant, and generous; an affectionate husband, a faithful friend, a father to the deserving soldier; gentle in his manners, in temper reserved; a total stranger to religious prejudices; in morals irreproachable; and never known to exceed the bounds of the most rigid temperance. In a word, all his friends and acquaintances allow that no man ever united in his own person a more perfect alliance of the virtues of a philosopher with the talents of a general. Candor, sincerity, affability, and simplicity seem to be the striking features of his character; and, when occasion offers, the power of displaying the most determined bravery and independence of spirit."[54]

All praise aside, George was still human—and thus susceptible to anger and frustration. One such instance was before the Battle at Monmouth in June of 1778. With Washington on one end and Henry Clinton and Charles Cornwallis on the other, the battle led to over one thousand casualties. It was a frustratingly complicated battle that ended up testing General Washington's resolve—and vocabulary. General Charles Scott of Virginia, a fellow Patriot participant, confided to a friend, "Never have I enjoyed such swearing before, or since. Sir, on that ever-memorable day he swore like an angel from Heaven."[55]

AS FOR MARY AT THIS TIME, HER AGE WAS A DETERRENT TO ANY ACTIVE role in the war, one way or another. She could not, like many wives, go on caravans with the army, as was popular during the later years of the war. The problem of hitching a ride with the men was so troublesome that Washington had issued around twenty-five orders regarding this. "Any woman found in a wagon contrary to this regulation is to be immediately turned out," read one. It was a source of frustration for the general.[56] His mother wouldn't go this route.

Her Fredericksburg house continued to be her place of residence, and whether she, in her age, went out to the bustling port city or stayed put inside, alone, was up to anyone's imagination.

George Custis believed her to still be a self-sufficient widow, doing her own gardening, her own duties, without the apparent need of neighbor, friend, or family member. Some inhabitants remembered her still visiting Ferry Farm, her place of residence for so many decades.[57]

Most recent historians have completely disregarded any hagiographical portrait of the woman, in which she was painted as the grandmother and redeemer of America. There was a certain romanticism in these works, which gave the impression that she was without sin. "The mothers and wives of brave men must be brave women," Mary supposedly said during the Revolution, according to Sara Pryor.[58] Yet some historians now go far in the opposite direction, painting her as almost the fiercest Loyalist in the colonies. Documents exist saying as such. The Comte de Clermont-Crevecoeur, Jean François Louis, a major French ally who visited Fredericksburg in July 1782, wrote that he and his party "went to call on [Mary] but were amazed to be told that this lady, who must be over seventy, is one of the most rabid Tories. Relations must be very strained between her and her son, who will always be the right arm of American freedom." The Comte's implication of Mary's beliefs was made even more clear; he noted that as amazing as it was, it wasn't uncommon: "In many families you find two brothers, or sometimes a father and a mother, holding opposite opinions. One is a defender of liberty, while the other is a confirmed Loyalist. What evils result from this division of opinion, which disturbs the union and sows discord in the midst of families who should be happy together!"[59] Another very early biography of Washington—published in 1807—said the mother of George was "so far from being partial to the American revolution that she frequently regretted the side her son had taken."[60]

Was Mary, the mother of George, a fierce Loyalist? The diary entry should close all matters. But Michelle Hamilton, manager of the Mary Washington House in Fredericksburg, shrugged off the Comte's diary. "I think we found the one person in town who didn't like the Washington family," she said.[61] Douglas Freeman—who

Hamilton said "made up some facts" about Mary—notes that Washington never really had affection for his mother, which he called "the strangest mystery of Washington's life."[62] Was it because of irreconcilable differences during the most important stage of the colonies, and their independence?

In reality, there was probably a grain of truth in the Comte's gossip. Mary was entrenched in the status quo. It makes sense that the persistent old matriarch would have been a royalist by default. However, given her lack of political activism, the Comte's claim that she was "one of the most rabid Tories" must surely be an overstatement. Mary's royalism was that of a small-T tory.

But in accounts of Mary's beliefs, the shift between completely loyal mother to "rabid" Loyalist Tory gained steam in the late nineteenth century, when funding and news of a monument dedicated in downtown Fredericksburg to her was hitting the presses. The whispers and gossip had always existed, however, much to the anger of some. Mary Terhune, who heard this view espoused more and more during her writing, called it a "rumor" and a "calumny."[63] She cited many biographers, some more reliable than others in modern histories, who personally knew her or knew someone who, in turn, personally knew her. For instance, George Custis, Mary's great-grandson, said it was an "absurdity of an idea which, from some strange cause or other, has been suggested, though certainly never believed." Writing in 1860, Custis unabashedly stated that no proof, "not the slightest foundation," existed for this claim. He did not believe that Mary was a Loyalist, nor, important to note, did he believe she was a Patriot. "Like many others, whose days of enthusiasm were in the wane, that lady doubted the prospects of success in the outset of the war, and long during its continuance feared that our means would be found inadequate. . . . Doubts like these were by no means confined to this Virginia matron."[64]

In reality, what loyalty she had to the Crown would have been more practical than ideological. For sixty years of her life, she lived under the royal banner of the king of England, under the

Parliament of Great Britain. For sixty years, she had accepted the king as monarch, as leader of both state and church. Why should that change? France and Spain had kings, as did Russia, as did the Holy Roman Empire, as did the Papal States. All of the European superpowers had royalty, so why shouldn't the colonies? There was no need for a revolution when she'd lived all her life, well into a mature age, without one.

"She probably was a Loyalist in the beginning of the war," Michelle Hamilton admitted. "I think she is on the fence. She is a woman. She didn't see this as her role in this society. Revolutions against the British Empire don't go well. She's living when she hears stories of the second Jacobite Rebellion, where all the Highlanders are wiped out."[65]

As the years went by, and this new discomfort became a way of life, Mary would have become more accepting of the situation. Certainly, her world was more involved in the war. On Christmas Day, 1776, her son crossed the frozen and dangerous Delaware River in a surprise raid on the German mercenaries in nearby Trenton. Tales say that Hugh Mercer himself, Mary's onetime physician and neighbor who bought Ferry Farm in what seemed like a lifetime ago, and who loved the fight more than medicine, suggested the perilous journey. The next day, Washington led several thousand troops to victory at the Battle of Trenton against the surprised Hessians with minimal casualties, though among the wounded was James Monroe, future fifth president of the United States.[66]

Mary would've been the perfect conduit between the south and the north, hearing news of her son's victories and failures. Said Benson Lossing: "Madam Washington was now in the direct line of communication . . . and she was in the constant receipt of news concerning the progress of the struggle at all points."[67] And though he then wrote on about a constant stream of couriers arriving at her house with letters from George (of which none are known to exist), women would have flocked to her, hoping for any word of their family and friends. Given the stature conferred on the mother of George Washington, no greater presence or woman of importance existed in Fredericksburg than Mary Washington.

IN MID-JANUARY 1777, SHE WOULD HAVE GOTTEN WORD, ALONG WITH ALL OF Fredericksburg, that neighbor Hugh Mercer, brigadier general of the Continental Army, was repeatedly bayoneted and stabbed at Princeton, New Jersey, succumbing to his wounds nine days later. Five days after his death, on January 17, the *Virginia Gazette* reported it with great dread, though they "hoped" it was not confirmed.

It was, however, very true. And the day before the *Gazette* published that piece, Mercer was interred with military honors in Philadelphia: with "the honours due to his merit, as a soldier, a patriot, and a friend to the rights of mankind. A numerous concourse," said the *Gazette*, "of people attended the awful solemnity."[68]

When Mary and the city heard of the news, it was a great loss. Mercer, as a physician and surgeon, had treated everyone in the region, from the well-off plantation owners to the dispensable slaves.[69] The town of Fredericksburg and the colonies mourned, as Mercer was a popular citizen of both. His importance to Fredericksburg never dwindled, either; in 1906 a larger-than-life bronze statue was erected in the town, depicting him with saber and hat in hand. But when Mary heard of his death, the man who very possibly, six years earlier, saved her life from an illness, was she feeling her son was next in line to lose his life? George had been there in that same battle.

Would George, her son, even survive?

DEATH WAS BUT ONE OPTION FOR HER SON. THERE WERE MANY OTHER POSsible conclusions, of course. It could have crossed Mary's mind: that her sons, her neighbors, her relatives both close and distant, were traitors. Traitors to the mightiest empire in the world, which had both the supremacy of the army and the navy and, some said, religion. Many knew that, recognized it, and embraced it. If they had failed, their penalty would have been the gruesome death of being hanged, drawn, and quartered. Some were only hanged, such as colonial spy Nathan Hale, whose last words in September 1776, "I only regret that I have but one life to lose for my country," were inspirational for the rebels in the early stages of the Revolution. But as the war continued, and more resentment grew in England, the punishments became even harsher. And George Washington, as the leader of these rebels—had he been captured, had he lost, had he made a command mistake—would have suffered an unimaginable death.

Perhaps he would have been whipped before death, very biblically. The whip and cat-o'-nine-tails striking his bare back and buttocks, each snap echoing. "At the first blow," a certain Private Alexander Somerville wrote decades later, "I felt an astonishing sensation between the shoulders under my neck, which went to my toenails in one direction, my finger-nails in another, and stung me to the heart as if a knife had gone through my body."[70]

WHATEVER HER PERSONAL THOUGHTS ON INDEPENDENCE FROM THE Crown, Mary still cared about her son and his safety. Some historians painted this as too controlling, as if a mother's fears for her son somehow dissipated after a certain age.

One story had been passed on through generations from her descendants. In historic Fredericksburg, now overlooking a tennis court and the sacred cemetery, right near her memorial, is Meditation Rock. This rock, sharply jutting out from the hillside, was one of Mary's favorite spots. She was keen on taking her grandchildren there, reading passages from the Bible, especially the creation of the universe and the world, the great flood of Noah, and other stories of the miracles of God in nature. "There was a spell over them as they looked into grandmother's uplifted face," said one of her grandchildren decades later, recalling that time in his own youth. There was a "sweet expression of perfect peace."[71] Here, Mary was at peace.

During the Revolution, that peace was disrupted across the colonies. One couldn't hide from it. But Mary tried to get away, somehow, from the thought of her son leading men to their deaths, the cause for freedom and independence or not. A nearby plaque at the spot summarized this importance for Mary: Here, during the Revolution, she prayed "for the safety of her son and country." Here, she spent hours, looking into the sunset each night.

And so Mary sat here, hands clasped, praising the Lord for the safety of her own son. Some previously called the site Oratory Rock, emphasizing the communication with God at this spot. Heaven opened up for Mary here. Here, at Meditation Rock, she got away from the world, from noisy Fredericksburg, from her home or her daughter's nearby plantation. Here she was with nature as God created it.

As Mary prayed on Meditation Rock, perhaps, at the same hour, her own son, hundreds of miles away, was likewise praying. It was this same faith she passed on to her children that was evidently vital to their lives. Mary Thompson, historian at Mount Vernon, believed there is strong evidence—both written and oral—that George Washington devoted time daily to pray in private.[72]

In April 1777, George, stationed in New Jersey, wrote to fellow Virginian and longtime correspondent Landon Carter. He hoped that "the God of Armies may Incline the Hearts of my American Brethren to support, and bestow sufficient Abilities on me to bring, the present contest to a speedy and happy conclusion, thereby enabling me to sink into sweet retirement, and the full enjoyment of that Peace & happiness wch will accompany a domestick Life is the first wish, & most fervent prayer of my Soul."[73]

AT THE CLOSE OF 1777, A WINTER STORM WAS THREATENING THE REBELS. The Continental Army had to seek safety near British-controlled Philadelphia, and on December 18 was ordered to build huts from fallen trees at Valley Forge. Huddled in the cold, twelve thousand men stayed until June of 1778.[74] Conditions became unbearable, with thousands losing their lives due to disease and frostbite. One of the men who survived was Henry Cone of Connecticut. About thirty-one years old at the start of the Revolution and at the time unmarried (he would marry Waitstill Champion in 1785), Cone immediately enrolled in the 1st Connecticut Regiment, participating in early major battles such as the Siege of Boston and the Battle of Bunker Hill. Later discharged, he next enlisted in the 3rd Connecticut Regiment on November 24, 1776. A year later, he was suffering at Valley Forge; again he survived. He and so many others had experienced their worst days here. (Cone, in 1781, lost an eye to smallpox, proving that even under the worst conditions, he would continue to fight.)[75]

IN THESE INTERVENING YEARS, FRIENDS AND ACQUAINTANCES WOULD VISIT Mary, staying and dining with her, as they inquired about George's whereabouts and safety and health. George himself was constantly "impeded" by all the crowds and audiences that wished to see him and touch him.[76] Such a devotion would have naturally spilled to the Fredericksburg mother. After the capture of Morristown, and the victory of the two battles at Trenton and the battle at Princeton by Washington, neighbors visited her to relay the news.

She replied not with outright praise or elation, but instead with unusual calm: "George appeared to have deserved well of his country for such signal services. But, my good sirs, here is too much flattery! Still, George will not forget the lessons I have taught him—he will not forget himself, though he is the subject of so much praise."[77] But with each victory came new recruits and volunteers and more support for Washington from the countryside and the congress.

SUCH STORIES AND TRADITIONS IN WHICH SHE CALMLY REPLIED TO NEIGH-bors celebrating the victory, sitting in her chair at her house, with barely any emotion, could do more harm than good. Historian Rupert Hughes, writing in 1930 in his multivolume work of the general, put it simply: "The denial of a trace of elation after such a crisis amounts almost to denying that Washington had a mother at all. It orphans him." Truly, such news would have been more than a simple hand wave dismissing the whole matter. Hughes argued that this description of an almost drone-like response, with no emotion, was about as biting as anything he had ever published.[78]

When George wasn't on her mind, her house and livelihood were. A smallpox epidemic continued ravaging the colonies, and in order to be immune, people were told to inoculate themselves. Fielding Lewis successfully did so, yet Betty contracted the disease, very mildly. Mary, however, did not want to be exposed. "Mrs Washington underg[oes] great uneasiness for fear she should take it," wrote George Weedon. "She cannot be persuaded to Innoculate, tho' it has been, and is now in almost every House in Town & Country."[79] At such an age, to be deliberately introduced to the disease would surely have been a death sentence. Mary thought so, at least, and whether Fielding and his wife convinced her to be inoculated is unknown.

If she was sick, she soon recovered and was in her old fighting spirits again. In December of 1778, she wrote a letter to Lund Washington, who was overseeing Mount Vernon ("Munt Vernon," Mary wrote) in George's absence, asking for 40 pounds of cash; the corn crop at her Little Falls quarter nearby had failed. "The plantation thear is terruable," she wrote. With corn at 5 pounds a barrel, she could not afford much. Within a couple of weeks, she received the money.[80]

The requests would soon come to a head.

THE YEAR WAS 1781: IT HAD BEEN SIX YEARS SINCE THE START OF THE REVO-lution, five since the Declaration of Independence. Battles had been won and lost; thousands of men had died, been wounded, been captured. Though the fight was almost over, there was still a pivotal port in Virginia—and that was Yorktown. A push in early 1781 from the British forces, led by Charles Cornwallis, into the Virginia Colony from the south, proved an unexpected success.

The war was again getting closer to home than Mary or George or Martha would have perhaps wanted. "My good son should not be so anxious about me," Mary was said to have declared when hearing of the advancing British, "for he is the one in danger, facing constant peril for our country's cause. I am safe enough; it is my part to suffer, and to feel, as I do, most anxious and apprehensive over him."[81]

That was oral tradition—legend—that had her report in such a selfless way. But in reality, there were more pressing matters for her, and that was her well-being.

She had been suffering, crops failing, her health declining. So she petitioned not her son, who was up north fighting, but the Virginia House of Delegates itself.

Her eldest son did not see it that way. He saw it as fiercely embarrassing. Worst of all, Mary did not tell her son she was going over his head. George had to learn it from someone else, from Benjamin Harrison, speaker of the Virginia House of Delegates, via a letter on February 25. She had either gone to or written the House some time earlier, pleading with them that she was out of money, and needed to pay her taxes. It almost came to a vote, but Harrison stopped the proceedings, noting that he had supposed George "would be displeased at such an application." He called the entire matter "an interesting subject."[82]

Interesting, indeed, but infuriating. George was irate. He had just returned to New Windsor, New York, from Hartford, to get word of this. The British advance into Virginia was bad enough; this was just as bad for an entirely different matter.

In a short but biting letter (he noted it was "written in much haste to go by the present mail, which is on the point of closing"), George wrote back immediately:

True it is, I am but little acquainted with her *present* situation, or distress, if she is under any. . . . Before I left Virginia, I answered all her calls for money; and since that period, have directed my Steward to do the same. Whence her distresses can arise therefore, I know not, never having received any complaint of his inattention or neglect on that head. . . . Confident

I am that she has not a child that would not divide the last sixpence to relieve her from *real* distress. This she has been repeatedly assured of by me: and all of us, I am certain, would feel much hurt, at having our mother a pensioner, while we had the means of supporting her; but in fact she has an ample income of her own.

In this same letter, perhaps cathartic to George, he wrote the many pains that he went through to get her comfortable in Fredericksburg, including the allowance, being so close to Betty and Fielding, and every other imaginable thing she may want. He continued,

I lament accordingly that your letter, which conveyed the first hint of this matter, did not come to my hands sooner; but I request, in pointed terms if the matter is now in agitation in your assembly, that all proceedings on it may be stopped, or in case of a decision in her favor, that it may be done away, and repealed at my request.[83]

In other words, "If she asks again—do nothing."

WHATEVER REPUTATION SHE HAD IN THE PUBLIC EYE, MARY DID STILL have friends and acquaintances in high places. Her most distinguished guest, perhaps the highlight of her life, was in April 1781. His full name was Marie-Joseph Paul Yves Roch Gilbert du Motier. More commonly, he was known as the Marquis de Lafayette.

Lafayette, the young grand French ally who had come to Virginia to assist colonial troops, purposely diverted his path to Fredericksburg to see Mary. While in Head-of-Elk, Maryland, he wrote to "my dear General" that "I could not resist the ardent desire I had of seeing your relations, and, above all, your mother, at Fredericksburg. For that purpose I went some miles out of my way, and, in order to consolidate my private happiness to duties of a public nature, I recovered by riding in the night those few hours which I had consecrated to my satisfaction."[84]

Such a difference in age, in customs, in beliefs. Here, a twenty-three-year-old dashing young man, French and European in culture, a veteran of wars and battles, an ally close to Mary's son at Valley Forge, sitting with a seventy-plus-year-old woman who had never left her Virginia Colony, who had been widowed for decades, and raised on a plantation. Here was a Frenchman whose friendship with her son was so great that he named his son Georges Washington Lafayette. General Washington returned the affection in kind by treating Lafayette as if he were his own son. Here was a man whose cause for liberty extended not just to the French but to the Americans. In many ways, they were sharp contrasts: young versus old, active versus deliberative, but their friendship and affection for each other knew no bounds.

Lafayette left within a day, heading back north to Maryland, leaving Mary pleased, no doubt, with his visit. Whatever conversations they had, it left an impression on the young fighter as well. "Be so kind, my dear general, to remember me to your much respected mother: her happiness I heartily partake of," he wrote to George in early 1783.[85]

WITH THE AID OF FRENCH AND SPANISH ALLIES, THE SIEGE OF YORKTOWN, months later into 1781, was the final blow to the British forces in America. Lord Cornwallis surrendered his forces, bringing any hope of vanquishing these rebels to an end. The Continental Army, commanded by George Washington himself, who lost more battles than he won, whose military training proved invaluable to the present crisis, had bested the greatest empire in the world. Cornwallis, bitter, did not present his sword in person to Washington. He sent his military aide; so Washington, tit for tat, sent his own military aide to accept it.

Hearing of Cornwallis's surrender, Mary was said to have clasped her hands in that ancient form of prayer, looked upward to God, and exclaimed, "Thank God! War will now be ended, and peace, independence and happiness bless our country!"[86] This demonstrated a marked change from Mary's earlier apathy. Years earlier, as she was gardening, a young courier announced, "Madam, there may have been a battle." He was greeted with a two-word reply: "Very well."[87]

Perhaps her praise at the surrendering British would mean an end to the constant couriers and messengers. Peace for the country, and peace and quiet for her.

GEORGE, HEADING BACK NORTH, DID NOT ARRIVE AT FREDERICKSBURG UN-til November 11, nearly a month after Cornwallis's capitulation. He paid her a visit, hoping to see her after all these years.

She was not at home.

Historian Douglas Freeman waved it all away: "He stopped in Fredericksburg to see his mother, but as she was absent on a visit, the party pushed on towards Mount Vernon." Ron Chernow provided some more detail, explaining that she was in western Virginia with Betty and Fielding Lewis.[88] Historian James Thomas Flexner only speculated on the move and motives. "I learn from very good authority that she is upon all occasions, and in all Companies complaining of the hardness of the times, of her wants and distresses," George wrote to younger brother John Augustine two years later, a complaint that Flexner felt weighed on Washington's mind while riding up to the Fredericksburg reunion. Ultimately, Flexner believed, George "was probably not heartbroken to find that his mother was not home."[89] Evidently while he was there, he dropped off 10 guineas before leaving. Mary wrote to him some months later sincerely yet dramatically apologetic. Dated March 13, 1782, to the misspelling of "my dear Georg," her letter said: "I was truly unesy [*sic*] by Not being at hom [*sic*] when you went tru fredirceksburg [*sic*] it was a onlucky [*sic*] thing for me now I am afraid I Never Shall have that pleasure agin [*sic*]."

It was her characteristic misspelling; that was nothing new and certainly nothing new for a woman in the 1780s, given the educational norms and habits of the time. Perhaps the misspelling of his own name warranted a quiet sigh from George, maybe a shake of the head in disbelief. But it was just like his mother. She thanked him for the "2 five ginnes you was soe kind to send me i am greatly obliged to you for it."[90]

It's important to note that Mary's absence should not be considered a snub to George. She was away from Fredericksburg with Betty and Fielding, not on some vacation or family-bonding "trip," as one wrote. She was not away because she wanted to be away from her war-distracted eldest son. She fled with Betty and Fielding from the potential colony-wide battlefield. Cornwallis was on the move toward Fredericksburg itself. The Marquis de Lafayette told General George Weedon, a Fredericksburg native, to "collect the militia" and evacuate all nonessential civilians. That included the Lewises and Mary.[91]

At the very north of Virginia, bordering Maryland, to which they fled, was Frederick County, created in 1743. Fredericksburg and Frederick County were both named for the then prince of Wales. George was the representative of the county for the House of Burgesses for several years.

The distance between Fredericksburg and Frederick County was about seventy miles, but these seventy miles were as far north as Mary would have ever gone, as much as can be known. The vast Shenandoah Valley, with the massive Allegheny and Blue Ridge Mountains stretching hundreds of miles away, offered a different view than that of the comparatively small Rappahannock River. But it was not an easy ride. "This trip over the Mountins [sic] has almost kill'd me," she wrote to George. She possibly stayed at Fairfield, the house of her distant relative Warner Washington. Despite the journey, she apparently loved it there.

Her time there may have been spent praying—or complaining, perhaps. For a woman of advanced age, as she was at this time, a perilous and uneasy journey like that would have taken its toll.

When George read her letter, he perhaps chuckled—or gasped, when he read that if she were to "be driven up this way agin [she] will goe in some little hous of my one if it is only twelve foot square[.] Benjamin Hardesty has four hinderd akers of Land of yours jis by George Le[e] if you will lett me goe there if I should be abliged to come over the Mountaine again."[92]

But that was an unnecessary request, even for her. And it was a request that was never answered. Some months after her son had passed through and with the threat of war over, she returned to Fredericksburg.

There, she resumed her life. She could meet her neighbors; she could shop and buy material she wanted or needed. On September 8, 1783, a mere five days after the formal end of the Revolutionary War, she received a bill from Captain Marban for 16 shillings for her purchase of two old blankets.[93] Why she needed them, or why she wanted them, who knew. But it was a sense of normalcy that she hadn't seen for the past eight years.

There, in Fredericksburg, to the day she died, she stayed.

ANY JOY OF VICTORY OVER THE BRITISH WAS DAMPENED BY THE DEATHS OF some very dear to her. For they did not die of their wounds or illness contracted from the war.

On September 26, 1781, at the age of forty-seven, her third child, her second son, Samuel, died at his home at Harewood, present West Virginia, built about twelve years earlier. Made of limestone, the Georgian manor was an oft-visited site of his older brother. It was unknown whether Mary herself visited, but that was unlikely due to her age. Samuel was the second child to pass away, after baby Mildred four decades earlier. He was the first of her adult children, which probably made it the most stinging. Samuel had followed the Washington "tradition" of dying early—his half brother Lawrence and father, Augustine, being testament to this.

Samuel was buried at Harewood, leaving behind at least five children.

Two months later, another heartache: Fielding Lewis, the husband of Betty, died of tuberculosis. He had moved away from his Fredericksburg mansion (completed in 1775) briefly in hopes of finding a cure, but returned soon after, dying in debt on December 7.

He had been sick, as many at that time knew, and was deep in debt from pouring his fortune into the munitions plant at Fredericksburg. At one point, he had advanced over 7,000 pounds for its manufacturing.[94] He had seen Mary's own health failing fast, offering to manage her affairs. She initially declined, but soon met halfway. "Do you, Fielding, keep my books in order . . . but leave the executive management to me."[95]

And now he was dead and had left Betty in debt. As a dearly loved member of her extended family, Mary saw the toll that this war had taken.

THE REVOLUTION, WHICH WOULD FORMALLY END AT THE SIGNING OF THE Treaty of Paris on September 3, 1783, was practically over. The surrender of Cornwallis's army was the end of any substantial fighting in the colonies, and British forces slowly retreated back to England.

The deed was done.

What was to be done with the Loyalists, those traitors to American freedom—yet those loyal to the Crown? "For the most part, open Loyalists had disappeared long ago." Those who left had all holdings confiscated. They fled to the West Indies, or England, or Canada.[96] Those who remained tried to save face, claiming that they were *actually* spying for the Americans. (In one case, for James Rivington, this was true.)[97] Mary, the supposed greatest Loyalist of all, as some may want to paint her, did not flee. Nor was she tarred and feathered or financially punished. In fact, quite the opposite. She remained home in Fredericksburg.

Before the end of the war, in mid-February of 1783, a baby girl was born of Burgess Ball, captain of the 5th Virginia Regiment and later colonel of the 1st, and his wife Frances Thornton Washington, daughter of Charles and granddaughter of Mary Ball Washington. She was the third child of the couple. In honor of her great-grandmother, she was christened Mary Washington Ball. This little Mary, like so many infants at the time, died within two years, on August 27, 1784, in Loudoun County, Virginia.[98]

All the while, from 1781 to 1783, George traveled around the colonies. He went to Philadelphia. He went to New York. Peace talks began in Paris in the spring of 1782 and continued on for the remainder of the year. Before he addressed the Continental Congress, on December 4, 1783, at Fraunces Tavern in New York City, he bid farewell to his generals. Colonel Benjamin Tallmadge wrote his account in 1830, several decades later, in which he writes of the incident:

The time now drew near when the Commander-in-Chief intended to leave this part of the country for his beloved retreat at Mount Vernon. On Tuesday, the 4th of December, it was made known to the officers then in New York, that Gen. Washington intended to commence his journey on that day. At 12 o'clock the officers repaired to Francis' Tavern, in Pearl Street, where Gen. Washington had appointed to meet them and to take his final leave of them. We had been assembled but a few moments, when His Excellency entered the room. His emotions, too strong to be concealed, seemed to be reciprocated by every officer present. After partaking of a slight refreshment, in almost breathless silence, the General filled his glass with wine, and turning to the officers, he said: "With a heart full of love and gratitude, I now take leave of you. I most devoutly wish that your latter days may be as prosperous and happy as your former ones have been glorious and honorable."

After the officers had taken a glass of wine, Gen. Washington said: "cannot come to each of you but shall feel obliged if each of you will come and take me by the hand."

Gen. Knox being nearest to him, turned to the Commander-in-Chief, who, suffused in tears, was incapable of utterance, but grasped his hand; when they embraced each other in silence. In the same affectionate manner, every officer in the room marched up to, kissed, and parted with his General-in-Chief. Such a scene of sorrow and weeping I had never before witnessed, and fondly hope I may never be called to witness again.[99]

Later that month, speaking to the Continental Congress in Annapolis, and to the shock of many, he resigned his commission.

He could finally go home to Martha, returning to the lovely mansion of Mount Vernon, so close to the Potomac River. There, he was determined, he would stay.

And in Fredericksburg, close to the Rappahannock, the now seventy-five-year-old mother would stay.

Chapter 11

A Separate Peace

THE UNITED STATES AND POSTWAR PEACE

1784–1786

———————

*"If such are the matrons in America, well
may she boast of illustrious sons!"*

———————

Though the war against the British was over, another battle would soon engage George Washington: the battle to get back to a normal life. That included, unfortunately, in some ways, his mother.

But fortunately, it also meant returning to his love, Martha Dandridge Washington.

The War for Independence was the first time, since their marriage in January of 1759, the two were separated for any length of time. His letter to her before leaving for his command showed great grief.

There was no such grief or lamenting for his own mother, however, either when he left for Mount Vernon in 1754 or when he went to war in 1775. This would not necessarily have been unusual considering he had his own wife and family. Perhaps a letter or more existed and became lost; perhaps not. He did return to domestic life, though, having sorely missed his wife and his land.

GEORGE HAD ALWAYS BEEN A PLANTER. "AGRICULTURE IS THE FAVORITE employment of General Washington," wrote his official biographer, David Humphreys.[1] James Thomas Flexner included a lengthy chapter in his third volume on Washington entitled, simply and humbly, "Farmer Washington."[2] "George perhaps saw himself as a simple farmer, away from the fame of war and revolution and independence, tending to his crops. Since his retirement in 1759, he shifted his life back, from soldier to farmer. He loved the tranquil calm that the Mount Vernon estate, that he received so long ago, brought him. Besides Martha, it was his true love, and as he wrote a decade after the war, 'No estate in United America is more pleasantly situated than this.'"[3]

During the war, George had placed Mount Vernon's eight-thousand-acre plantation in the hands and care of Lund Washington, his cousin. Five years George's junior, Lund kept the general apprised of all news of the plantation, good and bad. In 1781, he listed the seventeen slaves (fourteen men, three women) who escaped on board the sloop HMS *Savage*, commanded by British captain Richard Graves, who docked on April 12 on the Potomac River near Mount Vernon. There, slaves took flight.[4] George was furious at his cousin's failure, writing to Lund that same month: "It would have been a less painful circumstance to me, to have heard, that in consequence of your non compliance with their request, they had burnt my House, & laid the Plantation in Ruins."[5] The British also took off with some of the mansion's valuables.

Lund's management was overall a failure, a failure that George would deal with for years after he returned. To Henry Knox, he wrote in 1785 that Mount Vernon and all his business "require infinitely more attention than I have given, or can give, under present circumstances. They can no longer be neglected without involving my ruin."[6] To his sister's son, thirty-three-year-old Fielding Lewis Jr., he was even more blunt: "I made no money from my Estate during the nine years I was absent from it, and brought none home with me."[7]

All problems aside, George was finally going home to tend to his own fields. It was his refuge. It was his sanctum sanctorum. He had lived there for twenty-nine years, actually owning it for over twenty. Here he tended to the fields and gardens. Here in the

1750s he had expanded the initial one-story house to include a second story; in the early 1770s, before the Revolution and before the colonies were turned upside down, he expanded it again, with a two-story addition on the south end of the mansion and a similar, as yet incomplete, addition to the north. Here George built a larger gristmill to replace the previous one built under his father.

Here was home.

THE MOTHERLY ANNOYANCE FOLLOWING HIM FOR DECADES WOULD NOT GO away. It certainly did not go away in the middle of a war, and it didn't go away in 1783, either, when there was peace.

On January 16, 1783, when an armistice was all but official with Great Britain, George wrote to his brother John an impetuous rant about their mother, Mary. "I have lately received a letter from my Mother," he said, "in which she complains much of the knavery of the overseer at the little falls quarter." He went on, in a multipage tirade, in much the same way he did earlier to Benjamin Harrison. He continued, "That she can have no real wants, that may not be supplied I am sure of—imaginary wants are indefinite, & oftentimes insatiable, because they are boundless and always changing . . .

"It will not do to touch upon this subject in a letter to her," he concluded, before moving on to another subject, "and therefore I have avoided it."[8]

ON A BIGGER SCALE, AWAY FROM FAMILIAL TROUBLES AND DOMESTIC TRANquility, the independent country now known as the United States of America, separate from the monarch and parliament of England, was struggling. "The American war is over; but this far from being the case with the American revolution. On the contrary, nothing but the first act of the drama is closed," said Benjamin Rush, one of the signers of the Declaration of Independence, in a letter to Richard Price.

The United States of America was not, in practice, united, nor was it intended to be.

It was currently under the Articles of Confederation and Perpetual Union, unanimously approved by all thirteen states on March 1, 1781, at the Second Continental Congress. Thirteen articles long, this fiercely debated and ultimately failed system of government allowed a confederation of states, relatively autonomous from one another, to rule as they saw fit. It had a deliberately and, in hindsight, frustratingly decentralized rule. "The said States," it reads, "hereby severally enter into a firm league of friendship with each other, for their common defense, the security of their liberties, and their mutual and general welfare, binding themselves to assist each other, against all force offered to, or attacks made upon them, or any of them, on account of religion, sovereignty, trade, or any other pretense whatever."

The congress, with one vote per state, would have the power to only declare war and war-related logistics such as appointment of military officials; sign treaties; appoint a president of the congress (who would have only a one-year term); and a handful more of routine tasks. Most important, the congress did not have the power to tax; that was up to the states, proving that the taxation-without-representation frustration a decade earlier had not waned. The Framers were acutely aware of the perils of centralized power, after suffering for so long at the hands of Parliament and King George III.

From the start, the Articles were a doomed experiment.

PEACE HAD ARRIVED, AND GENERAL WASHINGTON BECAME, ONCE AGAIN, Citizen Washington. He was frequently visited and adorned with honors; many of the visitors were not even American. There were a "great number of foreigners who come to see him" at Mount Vernon, wrote French ambassador Anne-César de La Luzerne in April 1784. Family, too, visited him in northern Virginia, especially his nephews and nieces and cousins.[9] As much as he probably dreaded leaving his home, he knew that he needed to go back down south to see his mother.

There were several differing accounts of George and Mary

reuniting. George Washington Parke Custis specifically said they were reunited in late 1781, shortly after the siege of Yorktown.[10] Custis, of course, had a knack for confusing some dates; he knew the grand scheme, but as with Mary moving to Fredericksburg during the war and not before, he too got this wrong.

At some point, a gala known as the Peace Ball was organized to celebrate victory at the Battle of Yorktown. According to Custis, it was held on November 12, 1781. Others have said it was in December 1783 or possibly in early 1784.[11] Historian David Matteson, writing in 1941, believed it inappropriate and unlikely that it was November 1781. Days before the supposed ball, George visited Martha and her dying young son, John Parke Custis, who passed away on November 5. "Would he have been likely to approve of a brilliant social occasion so soon after this bereavement?" Matteson asked.[12]

More likely, if a ball happened, it happened after 1783, on a specific day in mid-February.

George had tried for a month but was stymied with each attempt to go to Fredericksburg. "The intemperence [*sic*] of the weather," in his words in a letter to Jacob Read, "has obliged me to postpone from one day to another." He understood, though, that the visit would not be a simple greeting, but something that he was obliged to do as a son and as a local hero for Fredericksburg.[13] He arrived within the day of the letter's dating, February 12.

While Washington was in Fredericksburg, presumably between February 12, when he arrived, and February 15, when he left, the people threw an impromptu celebration. When a courier ran over to Mary reporting of "his Excellency" soon arriving, she immediately perked up. "His Excellency!" she exclaimed, "Tell George I shall be glad to see him."[14] (Some report this story differently, with additional words from the mother. Instead of shock, it was condescending disbelief: "His Excellency! What nonsense!" she was rumored to have said. The two words surely changed the tone.)[15]

That evening, the people met at the town center. Allied French and Americans all crowded around to see the special guest, George

Washington, and his mother, Mary Ball Washington, as well. "The foreign officers were anxious to see the mother of their chief," Custis recalled. "They had heard indistinct rumors touching her remarkable life and character, but forming their judgements from European examples, they were prepared to expect in the mother, that glitter and show which would have been attached to the parents of the great, in the countries of the old world."

Instead, Mary walked out with her son clasping her arms, heading for the center of the audience. Her advanced age required her to use him as a cane, as she leaned on him. She was not dressed in jewels or endless flowing gowns befitting royalty, but a "very plain, yet becoming garb, worn by the Virginia lady of the old time."[16] (She was known to wear unusually normal clothing for both formal and everyday occasions. Recollected Nellie Parke Custis Lewis, the sister of George Washington Parke Custis and wife to Fielding and Betty Lewis's son Lawrence, "She was always remarkably plain in her dress—I do not believe she ever had, much less wore, a Diamond ring."[17])

How was it possible that this elderly, mostly plain lady was the mother of the great George Washington, the French must have thought. After all, he was a hero, Father of His Country, the American Cincinnatus? Her plainness, her humility here, as she slowly walked with her son, was discordant.

"If such are the matrons in America," yelled one Frenchman, "well may she boast of illustrious sons!"[18] Mary Terhune (who cited the ball occurring in 1781, not the likely 1784) believed that it was either Marshal Rochambeau or Admiral de Grasse.[19]

The mother and son danced; both were dancers and loved the activity as shouts of praise and joyous applause echoed the chambers for continuous peace in the continent.

By ten o'clock in the evening, while the celebrations were just starting, she looked up at her tall son, and said, quietly, "Come, George, it is time for old folks to be at home."[20]

And so she left the crowd.

While no ball was cited in the *Virginia Gazette*, the citizens of

Fredericksburg were certainly in awe of their former commander in chief when he visited. In a short but heartfelt letter on February 11, 1784, the mayor and citizens wrote to George Washington,[21]

Sir

While applauding millions were offering you their warmest congratulations on the blessings of Peace, and your safe return from the hazards of the Field, We The Mayor & Commonalty of the Corporation of Fredericksburg, were not wanting in Attachment and wishes to have joined in public testimonies of our Warmest gratitude & Affection, for your long and Meritorious Services in the Cause of Liberty; A Cause Sir, in which by your examples and exertions with the Aid of your gallant Army, The Virtuous Citizens of this Western World, are secured in freedom and Independance. And altho: you have laid aside your Official Character we cannot Omit, this first Opportunity you have given us, of presenting with unfeigned hearts, Our Sincere Congratulations on your safe return from the Noisy Clashing of Arms, to the Calm Walks of Domestic ease; and it affords us great joy, to see you Once more at the place which claims the Honor of your growing infancy, the Seat of your venerable and Amiable Parent & Worthy Relations. We want language to express the happiness we feel on this Occasion, and which cannot be surpassed, but by Superior Acts (if possible) of the Divine Favor.

May the great and Omnipotent Ruler of Human events, who in blessing to America hath Conducted you thro: so many dangers, continue his favor and protection, thro: the remainder of your life in the happy society of an Affectionate and gratefull people. I have the Honor to be (in behalf of the Corporation) with every sentiment of esteem & Respect Your Excellencys Most Obt & most Hble Servant

William McWilliams
Mayor

Once again, Mary hosted the Marquis de Lafayette at her home. Nearly three years had passed since their last visit, and these years had aged the woman considerably. "There, sir, is my grandmother," said one of Betty's children to the Marquis as he entered the garden that Mary was tending.

"Ah," Mary replied, seeing the young Frenchman, "Marquis, you see an old woman; but come in, I can make you welcome to my poor dwelling without the parade of changing my dress."

And so for the second time, they met. They spoke about George, they spoke about the war, they spoke about life in America. "I am not surprised at what George has done," Mary said in response to Lafayette's continuous praise, "for he was always a good boy."[22]

Here, according to folklore, she baked for the Marquis her now famous gingerbread recipe. The recipe was found among other papers at Kenmore in the 1920s, calling for orange zest, orange juice, butter, sugar, molasses, and a variety of spices.[23] Soon after, he was off again, departing to Mount Vernon.

The Marquis's admiration for the increasingly frail woman continued. "I have seen the only Roman matron living at this time," he said to friends.[24] The years went by afterward and the Marquis still asked George to give well wishes to "your respected mother" on multiple occasions, up to and including in 1788.[25]

WHILE IN FREDERICKSBURG, HE OF COURSE, AGAIN, DEVOTEDLY GAVE MARY money, this time 10 guineas on February 15, which was 10 pounds, 10 shillings.[26]

BEFORE GIVING HER THAT MONEY, HOWEVER, HE WAS INVITED TO DINNER with the mayor and city councilmen, around two o'clock. "Language is too weak to express the heart-felt joy that appeared in the countenances of a numerous and respectable number of Gentlemen, who had assembled on this happy occasion." A thirteen-gun salute accompanied the dinner, each to a special toast: to the thirteen states, to the congress, to King Louis XVI of France, to the army, to the American ambassadors in Europe, to a healthy economy, and so on. It was an altogether grand spectacle.[27]

George, honored, wrote in reply to the Fredericksburg people, thanking them immensely. This was his childhood town, and thus had significance to him, no matter how much he had missed Mount Vernon. In a letter dated in mid-February, addressed "to the Worshipful the Mayor and Commonality of the Corporation of Fredericksburgh [*sic*]," he wrote:[28]

Gentlemen,

With the greatest pleasure, I receive in the character of a private Citizen, the honor of your Address.

To a benevolent Providence, and the fortitude of a brave and virtuous army, supported by the general exertion of our common Country, I stand indebted for the plaudits you now bestow.

The reflection however, of having met the congratulating smiles and approbation of my Fellow-Citizens, for the part I have acted in the Cause of Liberty and Independence, cannot fail of adding pleasure to the other sweets of domestic life; and my sensibility of them is heightened by their coming from the respectable Inhabitants of the place of my growing Infancy and the honorable mention which is made of my revered Mother; by whose Maternal hand (early deprived of a Father) I was led to Manhood.

For the expressions of personal affection & attachment, and for your kind wishes for my future Welfare, I offer grateful thanks, and my sincere prayers for the happiness and prosperity of the Corporate Town of Fredericksburg.

Go: Washington

Both the festivity and the letter were reprinted in the *Virginia Gazette* (now, in post-Revolutionary Virginia, going by its full name *Virginia Gazette, or The American Advertiser,* published under James Hayes from Richmond) the following week.

The one line in particular—"The honorable mention which is made of my revered Mother; by whose Maternal hand (early deprived of a Father) I was led to Manhood"—has been quoted numerous times in early biographies as proof positive of Mary's continuous and, more important, positive influence. Certainly the context of their relationship during this writing, the legend of the Peace Ball, and the overall atmosphere of postwar jubilating America, lent credence to that.

That did not mean all her problems were absolved, though, nor did her son forget the headaches she gave him.

WHEN GEORGE LEFT FREDERICKSBURG AND RETURNED TO MOUNT VERNON, he went back to business. Martha, years earlier, had agreed to raise her orphaned nephews and nieces, a move that author Ron Chernow believed made Mount Vernon seem like a "small orphanage."[29]

Soon after he arrived at home, letters went a-flying to Daniel Boinod and Alexander Gaillard, two French booksellers in Philadelphia. In them, George praised the humanities, which he said was "a duty which every good Citizen owes to his Country." He requested several titles, all multivolume, for his personal library. All were in English, as he noted that he could not understand either the French or Latin texts available. Titles included *An Account of the New Northern Archipelago*, an eight-volume set by Jakob von Staehlin, on the discovery of a series of islands east of Russia and west of Alaska; *The New Pocket Dictionary of the French & English Language*, two volumes, by Thomas Nugent; and several other geographical and political histories of Europe.[30] Clearly Washington wanted to continue his intellectual pursuit, and a revolution was not going to stop it. By the end of his life, over 1,200 titles appeared in his personal library; reading was a habit he wanted to keep.

To Elias Boudinot, former president of the Continental Congress from November 1782 to November 1783, he wrote in reply, apologizing for the delay, as he was on "a visit to my aged Mother . . . which engaged me several days." He admitted that the return to "a peaceful abode, & the sweets of Domestic retirement" would never be forgotten amid all the celebrations and merriment.[31] He similarly wrote that same day to Elias's sister, Annis Boudinot, whose poems about the general resonated with him deeply: "It would be a pity indeed, My dear Madam, if the Muses should be restrained in you," he said.[32]

OTHERS WHO PASSED THROUGH OR JUST LIVED IN FREDERICKSBURG WOULD have seen her frequently. Her age alone—by 1783, an elderly seventy-five, especially for that time—would have solicited excitement at greeting the mother of America's greatest hero. One such man, neighbor William Simmes, described her in mid-1783 as "equally active & sprightly" as her daughter Betty. "She goes about the neighborhood to visit our quality on foot, with a cane in her hand & sometimes a Negro girl walking behind her in case of necessity."

He continued, that she "talks of George without the least pride or vanity. She will not keep any carriage, but a chair and two old family horses. . . . The front windows [of her house] are always shut & barred—for she delights to live in a little back room or two where I have seen her sitting at work with a slave to attend her—such is her taste."[33]

Still others recalled her with a distinct straw hat and red cloak—perhaps the same cloak her son had bought her decades earlier. If she had been tall like her son, her age made her shrink in size, a cane becoming more and more necessary as she hunched over.

And life continued on for her.

LEAVING THE TRANQUILITY OF MOUNT VERNON BEHIND, WASHINGTON made more trips to Fredericksburg again in 1785 and 1786. On April 29, 1785, the barometer read the air pressure in the mid-seventies with storms to the east, with "not very warm" weather. While heading to Richmond to assist the Dismal Swamp Company, which was managing and draining the Great Dismal Swamp in southern Virginia and northern North Carolina, Washington of course stopped for the night at Fredericksburg. He did not stay with his mother, instead sleeping at Kenmore, but before greeting his widowed sister, he paid a particular stop to Mary. He then went off the next day, dining with Alexander Spotswood, the husband of his half brother's daughter, in Spotsylvania County.[34]

ON MAY 1, GEORGE REACHED RICHMOND AND, THE NEXT DAY, HE DRAFTED resolutions to help jump-start the Dismal Swamp Company he had founded over two decades earlier. Only one page of their resolutions existed, but Washington, and others, agreed that the money raised could be put to better use: "Be put into the hands of some proper person, and such person be empowered to engage as many German, or other labourers at Baltimore, or any other part of this continent, as the money will procure." If it failed, then slaves—thought to be inferior in skill to the German or "Dutch" laborers—were to be used.[35]

When returning home after business was finished in Richmond, he again stopped by his mother's and sister's. He dined with Betty and "spent half an hour" with Mary before heading off, where he slept in a tavern near Stafford Courthouse in northern Virginia. No word was written on Mary's or Betty's conditions.[36]

MOUNT VERNON WAS PERHAPS NOT AS BUSTLING IN PRODUCTION AS HE wanted, but it was still a plantation to befit a Virginia gentleman. He had at least 216 slaves on his various properties, including at Mount Vernon, ranging from children to the elderly. The slave children alone numbered ninety-six. Here were waiters, housemaids, seamstresses, "almost blind" spinners. Among the most reliable was Billy Lee, also known as Will, his valet de chambre, whom George purchased in 1768 for over 61 pounds. Later that year, he acquired more by rent from a certain Mrs. Penelope French, and by the time of his death in 1799, he owned 317 men, women, and children.[37] More were owned through either indentured servitude or leased from neighbors.

George still loved the plantation. It could not be emphasized enough. "To see the work of one's own hands," he wrote in 1787, "fostered by care and attention, rising to maturity in a beautiful display of those advantages and ornaments which by the Combination of Nature and taste of the projector in the disposal of them is always regaling to the eye."[38]

With a new year of 1786 came more duties at the plantation. But he still could not forget his old ones. On April 24, a particularly rainy Monday, he headed down to Richmond for some legal issues on land deeds as sold to him by George Mercer. He paid a stop at Fredericksburg and "dined at my mothers [*sic*]" before again heading out, spending the night at nearby General Alexander Spotswood's residence.[39]

And even though some things continued to seem routine compared to less than a decade earlier, the fame and wonder that George continued to receive baffled him. The fame that his mother received likely baffled him even more. In May of 1786, he replied to a letter from the Marquis de Lafayette's wife, Marie-Adrienne. It was in answer to a letter she had sent him over a year earlier, in April 1785, in which she included a postscript: "Should I dare, beg you to pay my respects to your mother, I will certainly receive, an additional pleasure, in america, to present my self in person my respectful hommage." Clearly the Marquis had been telling his wife of all the grand people he saw and met in the United States, Mary Washington included. Whatever respect he had for her, he passed on. What an impression she must have made on the French couple.[40]

Now, here was George, in spring of 1786, writing from Mount Vernon in response:

"Of all the correspondencies [*sic*] with which I am honored," he opened, "none has given me more pleasure than yours." He apologized for the horrible delay in replying; as it turned out, the letter had been stuck at Bordeaux for "a considerable time" before being shipped off again to America. But the shipment took a "very circuitous" voyage, further delaying the reply.

He further apologized for being unable to visit France. He and Martha were now older, of course, and a visit would be too much. His explanation was poetic and almost bittersweet. Going into retirement again placed a sense of normalcy on him. "The noon-tide of life is now passed with Mrs Washington & myself, and all we have to do is to spend the evening of our days in tranquility, & glide gently down a stream which no human effort can ascend." Little did Washington know that his life stream would be ending later than he presumed.

"My Mother will receive the compliments you honor her with, as a flattering mark of your attention," he closed the letter, and "I shall have great pleasure in delivering them myself."[41]

"ALL WE HAVE TO DO IS TO SPEND THE EVENING OF OUR DAYS IN TRANQUIL-ity . . ." If those words were true for the fifty-four-year-old planter-turned-general-turned-planter, they were even truer for the elderly Mary. In 1786, she was in Fredericksburg, and while still small in comparison to the likes of Richmond, it continued to grow steadily. The road or path between Richmond and Fredericksburg, seeing increased traffic, was ordered by the Virginia Assembly to be straightened for more convenient accessibility according to a statute in October of 1786.[42]

MEANWHILE, RELIGIOUS HISTORY WAS CHANGING. IN RICHMOND, A PIECE OF legislation was passed that would become a hallmark of American and global liberty. First drafted in Fredericksburg, in 1777, the call for religious freedom was a cause that Thomas Jefferson fought for dearly. It was called the Virginia Statute for Religious Freedom, written by the deist Jefferson.

On January 19, 1786, after years-long fights both small and large, the "act for establishing religious freedom" was signed into law. "Whereas, Almighty God hath created the mind free," the preamble began. Yet government and church officials had believed it possible to govern people's religious beliefs, even demanding monetary tithes to the Church of England and the Church of England only.

"Be it enacted," it said, "by General Assembly that no man shall be compelled to frequent or support any religious worship, place, or ministry whatsoever, nor shall be enforced, restrained, molested, or burthened in his body or goods, nor shall otherwise suffer on account of his religious opinions or belief, but that all men shall be free to profess, and by argument to maintain, their opinions in matters of Religion, and that the same shall in no wise diminish, enlarge or affect their civil capacities."[43]

The Church of England was no longer state sanctioned.

In the past couple of years, the official Anglican Church in the colonies had been going through an identity crisis. Anglican theology's whole philosophy, settled centuries earlier under King Henry VIII, asserted that the monarch of England, be it a king or queen, was the head of the institutional church, not some foreign pope in Rome. "Defender of the Faith" was and continues to be one of the formal titles of the British monarch.

The Americans had a dilemma: The War of Independence refused to recognize the Crown as the political and earthly leader. What was to happen to the Crown as its spiritual leader? The Revolution itself could have been classified as a holy war, changing the very fabric of the country not just socially or politically but religiously. George himself demanded that chaplains were needed throughout the Continental Army, the most impressive of which to him was Abiel Leonard of Connecticut. John Adams called the Bible "the most Republican book" he had read, inspiring many for truths. Thomas Paine, who was no friend of Christianity, still quoted from the Bible in *Common Sense*.[44]

The Episcopal Church, breaking away from the Church of England, was established. Its first bishop, Samuel Seabury, was consecrated in Scotland.

Mary and George did not really care for the exact doctrinal differences between Episcopalianism and Anglicanism. George resigned from the vestry of Truro Parish in 1784, a move that Mary Thompson, historian at Mount Vernon, believed indicated that Washington "could no longer promise to be subject to the doctrine and discipline of the Anglican Church" after leading such a grand rebellion. Pohick Church in Lorton, Virginia, of which his father was vestryman, was expanded in 1767. Washington had surveyed the land and contributed to its finances and construction. To resign from its vestry was not a small move.[45] To Mary, whose faith continued steadfastly, it would not make any difference, certainly. She would continue to go to Saint George's Church in Fredericksburg whether it was Anglican or Episcopalian. Theologically, it was the same. *The Book of Common Prayer* was similarly retained by both churches.

AS 1786 CAME TO A CLOSE, SO DID ANOTHER CHAPTER OF AMERICA'S LIFE. The year 1787 would be unlike 1786, unlike 1785, unlike 1784. The country had been going through creative turmoil. As Washington tried to adjust to normalcy, his nation would be plagued by rebellions, exploited weaknesses as befitting a flawed confederacy, debt, and incompetence. The Annapolis Convention tried to change that, but to little real immediate effect. By the end of 1786, the Seventh Confederation Congress was under way, again returning to New York. (Philadelphia, Princeton, Annapolis, and Trenton had all been previous locations.) The Northwest Territory, including the Ohio Valley that had been fought over decades earlier, was again the center of tension as the United States claimed ownership of Indian-controlled lands.

Would the country he desperately fought for, and so many men died for, itself fail? Was that his destiny?

Mary, in Fredericksburg, perhaps did not care. Or perhaps she didn't even know. Political misgivings were never her forte. And she wouldn't need to know or care. She lived all her life worrying only about her son's safety. Since there was no war, there was no issue requiring her to stay up at night. She would tend to her garden and live on.

The year 1787 would be destined to be a year of a second revolution, in ways different from the first a decade earlier. Instead of a bloody revolution, it would be a political one, a legislative one.

In 1787, this new convention would convene to draft the Constitution of the United States of America. Quite literally it was a new chapter in the country's history.

Mary, too, entered a new phase of life. It was her last.

As her country was beginning a rebirth, Mary was dying.

Chapter 12

Honored Madam

THE LAST YEARS OF MARY BALL WASHINGTON

1787–1788

"Age is a Disease, and the decay of nature . . ."

I am an old almanack quit[e] out of date."[1] So Mary Ball Washington wrote touchingly to her son John Augustine around 1781. Just eight years later, that old "almanack" would go out of print.

By 1787, Mary was at the remarkable age of seventy-nine. Life expectancy was short in colonial America, generally around the midfifties, though the statistic was skewed considerably due to infant and child mortality throughout that time period. Despite that, it was higher in the 1700s than in the 1600s, and historian Daniel Blake Smith, who studied this data, conceded that giving simple raw numbers "is too extreme because it indicates" that people died in their fifties and no older. Each stage of life marked a milestone in survival; if one survived to twenty, some Virginia men were expected to live another 28.8 years. If they reached fifty, they were expected to live another 7.7 years. And so on.[2]

Mary's age was somewhat unusual for the time, but not unique throughout the colonies. Benjamin Franklin, for instance, by 1787, was eighty-one, and would go on to live another three years. Thomas Jefferson lived to be eighty-three. John Adams lived to be ninety.

This was no small feat, either. For Mary, disease and childbirth could have taken her at any time.

THE YEAR 1787 WAS NOT ONLY IMPORTANT FOR THE GRAND THIRTEEN United States, but for George Washington himself. Early February, he laid down the law against his mother.

It was a disturbing point for him. Only a month earlier, in January, he lost his brother John Augustine. Mary had now outlived three of her children, including baby Mildred so many years before. John was about fifty-one years of age, near the same age at which his own father had died. For a third time, Mary was to go about her days knowing that another of her children had died before her, an event no mother would want to experience. This was the beloved "Johnne," as she affectionately called him, her fourth child. Only George, Betty, and Charles were now left of the original six.

Prior to his death, in an undated letter, she had written to John Augustine a plea for help: "I am a going [*sic*] fast," she wrote, "and it be time as hear I am borring a Little Corn no Corn in the Cornhous." She continued, "I Never Lived soe por in my Life [and] was it not for Mrs. French and your Sister Lewis I should be almost starvd."[3] It was a "legitimate" plea, according to Michelle Hamilton, manager of the Mary Washington House, as Virginia that year endured a particularly bad bout of crops that even hit George.[4]

Now, with John Augustine, the "mediator" of the relationship, as historian James Flexner called him, gone, there were "open fireworks."[5] These fireworks were often more like cannon fire, though.

George wrote an emotional letter.

After saying that 15 guineas would be delivered to Mary via John Dandridge, Martha's nephew, Washington got to the heart of the matter. It was poignant. According to her son, Mary had asked George Augustine Washington, her grandson and the son of Charles and nephew of the former general, for money, in a lost letter that read presumably as many of her other letters.

"I have now demands upon me," he said in a lengthy and angry note to Mary, dated February 15, "for more than 500£[,] three hundred and forty odd of which is due for the tax of 1786; and I know not where, or when I shall receive one shilling with which to pay it."

Mount Vernon over the past year had been a failure, with no wheat or tobacco or any sort of crop coming in. The corn Washington had grown that year was of horrible quality, and even he admitted that he found the taste repulsive. He asked others to pay him back money that they owed, and, in his view, they refused, taunting him to take them to court, which was "like doing nothing."

How was he to pay the taxes? How was he to live?

He blamed her.

"It is really hard upon me when you have taken every thing you wanted from the Plantation by which money could be raised— When I have not received one farthing, directly nor indirectly from the place for more than twelve years if ever—and when, in that time I have paid, as appears by Mr Lund Washingtons account against me (during my absence) Two hundred and Sixty odd pounds, and by my own account Fifty odd pounds out of my own Pocket to you." Even the Little Falls Quarter plantation on the Rappahannock, legally Mary's but leased to George, saw all its profits going to the elderly woman.

"I am viewed," he said, "as a delinquent[,] & considered perhaps by the world as unjust and undutiful Son." Word traveled fast when a son supposedly neglected his mother.

Because of the guilt she had apparently imposed on her eldest son, which he clearly saw now as a final straw after many years and decades, he offered her some advice: sell or rent out your slaves, sell the land, move in with "one of your Children." (Washington, when writing this down, perhaps was hoping she wouldn't pick him, hoping that Mary, if she took the advice, would pick Betty or John; he had previously spoken to John about it, but it wasn't to be.) Moving to Mount Vernon, though, would do no good; people came and went, and she would be obliged to greet each and every one, every day, and the relative isolation of the plantation would have her feeling trapped. Moving with Charles or Betty "would relieve you entirely from the cares of this world, and leave your mind at ease to reflect, undisturbedly on that which aught [*sic*] to come."

This, he continued, was "the only means by which you can be happy."

He closed the letter, "I am honored Madam, Yr most dutiful and affe[ctionate] Son, G. Washington."[6]

She did not take his advice. She never left her Fredericksburg house.

A WEEK BEFORE THAT FATEFUL LETTER TO MARY, THE MARQUIS DE LAFAY-ette again wrote to his American friend-in-arms. It concerned French international relations with Russia and other such matters. He ended it: "Adieu, My Most Beloved General, Be pleased to present My Best Respects to Mrs Washington, Mrs Stuart, Your Respected Mother, all your family."[7]

ASIDE FROM FINANCIAL AND FAMILIAL WOES, THERE WAS OTHER STRESS IN Washington's life. In late 1786, he had learned he was to represent Virginia, along with six others, at the upcoming Philadelphia convention. He had gone so far as to decline his appointment, and James Madison, ever vigilant, scolded him. "My name," George wrote to John Jay, "was put there contrary to my desire and remains there contrary to my desire." Friends on both sides of the debate—some demanding he attend, some cautioning its legal nature—weighed in, and so Washington deliberated until March 28, when he wrote to Governor Edmund Randolph.

He accepted his appointment.[8]

MARY WASHINGTON, APPROACHING EIGHTY YEARS, WAS GROWING FRAIL.
Certainly the cane she carried was evidence of weakened knees
and bones. The lack of children in her company perhaps made her
more bitter to youth.

She probably found comfort in either her beloved Bible or her
ever-reliable *Contemplations, Moral and Divine*, by Sir Matthew
Hale, a work she had had for decades. (A copy of *Contemplations*
sits in the Mary Washington House in Fredericksburg; another is
at Mount Vernon.)

Contemplations delves into Christ, morals, religion, faith. The
fear of God, the mediation of the Lord's Prayer, and how to keep
Sunday holy are all featured in the work. And old age was often
considered as well. When Mary read the book in the 1780s, as
she continued to age, she would read Hale's advice, meditating on
what he said. To Hale, old age, in his own words, "is a Disease,
and the decay of nature, heat and moisture, doth in time bring
the oldest Man to his end."[9] Everyone dies, everyone ages. In the
Garden of Eden, of course, humanity was physically immortal,
with no disease or illness or frailty. But after the Fall, humanity
was resigned to a more pessimistic memento mori.

According to Hale, Mary was to be humble and not to brag
about having wit, memory, prudence, or intelligence. While God
gave her all she had, she should not brag. "Thou art but a temporary owner of them," he wrote. "If thou live to old age (a thing
that naturally all Men desire) that will abate, if not wholly anti-
quate, thy Wit, Learning, Parts; and it is a foolish thing for a Man
to be proud of that which he is not sure to keep while he lives."[10]
The Book of Proverbs said clearly of pride: "Pride goeth before
destruction, and a haughty spirit before a fall." Mary's faculties,
whatever she may have had as she read the book, would soon be
gone. She should not be proud of that which she would lose.

He continued, many pages later, a point he wanted to emphasize, "How brittle and unstable a thing thy Wits, thy Parts, thy
Learning is. Though old Age may retain some broken moments
of thy Wit and Learning thou once hadst, yet the floridness and

vigor of it must then decay and gradually wither, till very old Age make thee a Child again, if thou live to it." Disease may strike, he said, at any moment, rendering your pride "under [its] mercy."[11]

Similarly, worldly possessions and worldly enjoyment will diminish. Friends will die. People will leave. You will be alone. "In the tract of long Life, a Man is sure to meet with more Sickness, more Crosses, more loss of Friends and Relations, and over-lives the greatest part of his external comforts, and in old Age becomes his own Burthen."[12]

Mary understood this clearly. Three of her children had died already. Her husband was long gone, as well as her mother and father. Her son-in-law Fielding too. Her neighbor and doctor Hugh Mercer was killed a decade earlier. Anyone she knew from her first years at Ferry Farm would be long gone. And as she continued to live, so there would be more.

MARY'S OTHER FAVORITE BOOK, THE BIBLE, OFFERED HER SOME COMFORT IN her aged years—comfort from Hale's depressing words.

When Mary looked in her small mirror and saw the gray hair, she may have thought of the Book of Proverbs: "The hoary head is a crown of glory, if it be found in the way of righteousness."

But there was still one goal in her life as she aged. And it was just as true when she was young as it was now: to be faithful for God and the proclamation in both words and deeds of His existence. Her attendance at Saint George's Church continued well past her prime, and she certainly saw her age just as much a proof of God as any other.

But those years, she noticed, would be coming to an end soon.

A MERE SHORT MONTH BEFORE THE OFFICIAL START OF THE CONSTITU-tional Convention, designed initially to correct the errors of the Articles of Confederation and result in the creation of a unique government, George Washington was in a state of unease.

If the stress wasn't enough, between four and five o'clock in the afternoon of April 26, a hurried and exhausted courier rushed to his Mount Vernon plantation with news of his Fredericksburg family. His mother and sister, the letter said, were in "extreme illness." The timing could not have been worse, as Washington was ready to head north to attend the convention. To boot, he himself was ill, suffering from rheumatism. He was to leave in three days, on a Monday, when the news arrived. In a letter to Henry Knox, he wrote that he was "summoned by an express who assures me not a moment is to be lost, to see a mother, and *only* Sister (who are supposed to be in the agonies of death) expire. I am hastening to obey this melancholy call, after having just bid an eternal farewell to a much loved Brother who was the intimate companion of my youth and the most affectionate friend of my ripened age." The trip to Fredericksburg would undoubtedly delay his going to Philadelphia.[13]

His life seemed to be coming apart all at once.

A fatal illness was almost expected for his mother at this point, given her advanced age. But his lovely sister Betty? At nearly fifty-four years old, she was still a woman of vigor. And now, she was dying along with their mother.

George traversed to Fredericksburg the next morning, a Friday, and arrived that afternoon. He "was prepared to expect, the last adieu to an honoured parent, and an affectionate Sister."[14] There, he was greeted with some good news. He found, he wrote in his diary, "both my Mother & Sister better than I expected." Betty was "out of danger" of whatever had afflicted her. But Mary had been diagnosed with breast cancer. She would live another two years, but at the moment, to George's relief, she was doing well. (Descendants of the Ball family reported centuries later that the women were susceptible to cancer, especially later in life.[15]) He knew there "was left little hope of her recovery."[16] But his fears of this being his final sight of either his mother or his sister were alleviated.

He spent the next couple of days in Fredericksburg, dining with his sister and neighbors. He left on Monday, April 30, returning to Mount Vernon.

Within a couple of weeks, he left his home, once more returning to public life. His wife, Martha, was again left at Mount Vernon.

He arrived at Philadelphia on May 13, escorted by a parade of soldiers and other officials. Many along the damp roads came out to see, as the *Pennsylvania Packet* called him "this great and good man."[17]

THE CONVENTION OFFICIALLY CONVENED ON FRIDAY, MAY 25, 1787, AFTER representatives from New Jersey arrived to form the necessary quorum of seven states. They met at the Pennsylvania State House, where a decade earlier the Declaration of Independence was signed.

The convention would last four months. In total, there were fifty-five delegates from twelve of the thirteen states, though not all fifty-five were there from the beginning to the end, nor were fifty-five delegates ever in the same room at the same time; Rhode Island, with a penchant for trouble in the newly forming United States, refused to participate. The previous years of trouble over the state's sovereignty from its neighbors earned it the irritating nickname "Rogue Island." From Virginia came seven delegates, including Washington, James Madison, George Mason, and Edmund Randolph.

The very day they convened, George's responsibilities at the convention expanded greatly: he was to be presiding officer of the convention. It was a unanimous vote.[18] Joseph Ellis describes his relationship with fellow delegates: "His importance was a function of his presence, which lent an air of legitimacy to the proceedings that otherwise might have been criticized as extralegal, if not a coup d'état." Note that the convention was called to amend, not overthrow, the Articles of Confederation, a distinct difference. Washington hardly if ever participated in the months-long and often fierce debates at the State House. As president, he was to lead and keep order, not to butt in. He'd leave the political philosophy to Madison and Benjamin Franklin and Alexander Hamilton and many other heavyweight thinkers.[19]

The discussion quickly evolved from "how to fix the Articles of Confederation" to "how to overrule the Articles of Confederation." It was clear to many that the Articles were not working, that they had become a failed experiment threatening the infant nation. Various plans were drafted and rejected or modified to create the final product known as the Constitution of the United States.

"The Constitutional Convention," wrote historian James Thomas Flexner, "was, in its most creative aspects, less a forum of debate than experimental laboratory."[20] Indeed, it was decided that the system of government, unlike any other that had existed at that time, did not work. How were they to replace it? With something familiar, with something different, with something wholly unique to history?

Three major plans were proposed:

One of most important was the so-called Large State Plan, or the Virginia Plan, the brainchild of Virginians such as Madison and presented to the convention on May 29 by Governor Randolph. George and the other Virginian delegates attended the brainstorming sessions days prior.[21] The Virginia Plan proposed the "National Legislature," which under the Articles was unicameral, be split into two branches and be representative of a state's population. It also asked that "members of the first branch" (which would become the House of Representatives) be elected directly by the people for fixed terms and then be ineligible, for a period of time, for other government positions after they finished their term.

The "second branch," the precursor to the Senate, was proposed to be elected by the first, with nominations coming directly from the states' individual legislations. Like the pre-House, the pre-Senate would be proportional, with fixed terms; those elected would be ineligible for other positions for a certain number of years.

The legislature would have the responsibility of electing the "National Executive," a separate branch from the legislature, for an unspecified length of time for only one term. That individual would have the power to "compose a Council of revision with authority to examine every act the National Legislature before it shall operate." This council, which had no specificity of how, when, or where it would operate, would then have the power to veto both national and state laws.

The Virginia Plan also called for a separate judiciary branch. Like the executive powers, these judicial appointments were a brand-new confederate system of government.[22]

Up next was the so-called New Jersey Plan, or the Paterson Plan after framer William Paterson. Keeping the Articles' unicameral congress, it provided additional powers of raising and collecting taxes, and placing tariffs, among other regulations. A federal executive of several people would be elected for one term, which would itself have the power to use the legislative-raised army and appoint the judicial "supreme tribunal," a lifetime position. While, again, executive and judicial branches were new, the New Jersey Plan more or less improved the roles of the one-house legislature of the Articles of Confederation, with each member nonproportionally representing each state, and each state receiving one vote. Madison was skeptical of such a plan, wondering whether it would prevent state-level abuses that so plagued the current crisis. "Will it secure the internal tranquility of the States themselves?" he asked. "Will it secure a good internal legislation & administration to the particular States? . . . Will it secure the Union [against] the influence of foreign powers over its members[?]"

In short, Madison believed, it would fix nothing.[23]

Alexander Hamilton of New York, later the first secretary of the Treasury, liked neither Randolph's nor Paterson's proposals. His own, called the British Plan, was very similar to the British government setup, in which Hamilton effectively did away with any statehood under the federal system. It called for a two-house Supreme Legislature: the Assembly, directly elected by the people; and the Senate, chosen by "Electors" who themselves were elected by the people. The Assembly would have three-year term limits, and the Senate would be elected for life. The Senate could declare war, along with advising and approving anything in the executive. The directly elected Assembly, on the other hand, had the power only to create laws and appoint state-level courts, with the Senate's cooperation. It was clear which house was "lower" and "higher" in Hamilton's plan.

These Electors, who were not chosen by the state but rather by "election districts," would also have the power to appoint a life-serving governor. The governor was effectively like the monarchy of Britain, and had a plethora of powers. This individual had the absolute power to veto any law passed by the Supreme Legislature. He was to command the army and navy forces, and have the power, "with the advice and approbation of the Senate," to make treaties. He was to nominate, again with the approval of the Senate, any ambassadors and "heads or chief officers" of the departments of war, finance, and foreign affairs. He also had the power to pardon any criminal, except for those convicted of treason, which needed Senate approval. When the governor died, the power would temporarily shift to the president of the Senate.

The similarity to the British system was not lost on anyone. Madison noted, "It would be objected probably that such an executive will be an elective monarch and will give birth to the tumults which characterize that form of gov[ernmen]t."[24]

THROUGHOUT, GEORGE WASHINGTON SAT AND LISTENED. THE LETTERS HE wrote were unrelated to the convention or his thoughts on the very future of the United States. He wrote to his manager of Mount Vernon, his nephew, George Augustine Washington, who had replaced Lund after the war. On June 3, the elder George wrote his nephew about observations from the plantation's previous harvest, in hopes that George Augustine would know what to do in the coming months. "All the grass that is fit for Hay should be cut, or my horses [etc] will be in a bad box next winter," he advised. "In making Bricks let the Mortar be well neaded [sic]—much

I believe depends upon it." The carrots and parsnips had failed, according to the estate's manager, and George Washington wondered whether it was bad seed or bad soil or bad weather, offering advice on the potato harvest. It was, all in all, a very plain letter from one farmer to another.[25] The letters continued as such for the next couple of months, noting expansions to the plantation house and the management of the logistical nightmare that was his home.

To fellow delegates, however, he had strong opinions. "In a word, I *almost* despair . . . and do therefore repent having had any agency in the business," he wrote to Hamilton, who had returned to New York soon after presenting his proposal. "The Men who oppose a strong & energetic government are, in my opinion, narrow minded politicians, or are under the influence of local views."[26] Insults were being thrown, distrust running rampant not just between cross-border delegates but among states' own. Some threatened to leave and never return, putting the already fragile convention at risk.

It was decided ultimately by the delegates of the Constitutional Convention to use the Virginia Plan as the basis for the new form of government. The so-called Great Compromise, first proposed by delegate Roger Sherman and subsequently modified, alleviated fears from smaller states that proportional representation would naturally favor the more populated states.

On July 24, a committee of detail was established to draft a rough outline of the to-be-formed Constitution. Chaired by John Rutledge of South Carolina and containing Virginians such as Randolph and Connecticut's Oliver Ellsworth, it created positions and institutions of "president," "Senate," "House of Representatives," and "Supreme Court." A two-house legislature, one with proportion and the other with a fixed rate per state, placated both the smaller and larger populated states. Twenty-three articles were drafted, making a rough if not workable document. Flexner noted the miraculous nature of the committee: "Who, for instance, could three months before have guessed that delegates from the north, the middle, and the south, from small states and large, would agree that the federal government could bypass all state bodies and collect taxes directly from individuals in the local streets and fields?"[27] The powers and limitations, both extraordinary, of all three branches proposed made the job of the next committee, the committee of style, that much easier. By September 12, it presented the final draft, with the twenty-three articles combined to a mere seven, with an additional preamble and eschatocol. A vote was raised on September 15, and it received unanimity from each of the twelve attending states. Two days later, thirty-nine of the forty-three attending delegates signed; George was the first. Edmund Randolph and George Mason did not sign, nor did future vice president Elbridge Gerry of Massachusetts. They all objected to a lack of guaranteed rights.

Washington's diary plainly stated: "The business being thus closed, the Members adjourned to the City Tavern, dined together and took a cordial leave of each other—after which I returned to my lodgings—did some business with, and received the papers from the secretary of the convention, and retired to meditate on the momentous wk. which had been executed, after not less than five, for a large part of the time Six, and sometimes 7 hours sitting every day, sundays & the ten days adjournment to give a Comee. opportunity & time to arrange the business for more than four Months."[28]

He went home, again, on September 18—"I am quite homesick," he wrote a week earlier to George Augustine—and arrived on Saturday the twenty-second, around sunset, "after an absence of four Months and 14 days."[29]

And though the fight for the drafting and approval of the Constitution was over in Philadelphia, it would move through the next year for ratification by each state.

NEARLY TWO HUNDRED MILES TO THE SOUTHWEST, MARY'S OWN FREDERicksburg was going through a miniature crisis of its own. Back on November 5, 1781, only a month after Lord Cornwallis surrendered at the Battle of Yorktown, it had been incorporated into a town, with the creation of its own mayor, recorder, alderman, and councilmen, twelve in total, as well as a court of law, all separate from Spotsylvania. They had the power to buy and sell land, to act as justices of the peace, to hold two "market days, the one on Wednesday, the other on Saturday, in every week, and from time to time to appoint a clerk of the market, who shall have assize of bread, wine, and other things."[30]

The residents of Fredericksburg apparently didn't like it. John Dawson, representative of the Virginia House of Delegates of Spotsylvania County and the stepson of James Monroe's uncle, was a vocal opponent. "You," he wrote to James Madison in June 1787, "I know, are oppos'd to the plan of incorporating towns, which in this state, has been so much in vogue, for some years past." He continued, "The people in this county, convinc'd of the bad policy, intend to petition the next assembly for a repeal of the law incorporating this town. I have promisd to forward their wishes."[31] Evidently, he overestimated the residents' anger at incorporation. The repeal, if it was ever submitted, failed; Fredericksburg was eventually incorporated as a city a century later in 1879.

George Washington remained composed yet concerned about his mother's illnesses. He still received reminders of her hometown: in September of that year, he received letters from Fredericksburg residents such as William Roberts, a miller George had once employed.[32]

MARY BALL WASHINGTON, AT THIS POINT, WAS PRETTY MUCH SETTLING into her final years. Fredericksburg grew around her; she stayed put. There is little evidence and only a handful of legal papers that open a window into her final years. Perhaps most people had more important things to do than speak with an elderly lady, even if she was the mother of the commander in chief.

MARY STAYED IN HER HOME OF TWO LOTS, VALUED AT 300 POUNDS. SHE REmained uninvolved with her son's role in Philadelphia, Mount

Vernon, or the events in Fredericksburg. Anything more about these months and last years of her life would be speculation. Sadly now deteriorating from age and cancer, Mary tended to her gardens, moving about the house with the help of a slave holding her arm. She sat outside and watched the young Fredericksburg children play in the snow of winter or the heat of summer. She feebly got on her knees as she did her daily prayer, or read Hale's *Contemplations* with an earnestness not previously found in her youth. Mary may have gone into town; she may have spent the remainder of her days locked in her house on Charles Street. One resident recollected decades later that she "was remarkable for taking good care of her ducks and chickens," so reported historian Jared Sparks as he visited the town in the mid-1800s.[33]

Tax records from 1787 indicated she had six slaves over sixteen years of age, three horses, one cow, and a four-wheeled open carriage. This was a decrease in her slaves in 1785 from ten total (five above sixteen years old, five below), though the next year, 1788, records report her losing one, bringing it down to only five slaves.[34] Still, that number of slaves was "unusually large" and "too many," opined J. Travis Walker, archivist at the Fredericksburg Circuit Court, for a house of its size and a woman of her age, and records are scant on whether she lent the slaves out for others' use.[35] It showed either a lack of management skills—or at the very least a deterioration of them—or a deliberate desire to keep things "the old way" as at Ferry Farm, where she was once in charge of their world.

THIS NORMALCY IN FREDERICKSBURG CONTINUED INTO 1788, DETACHED from the grander US politics. By New Year's Day in 1788, Delaware, Pennsylvania, and New Jersey ratified the drafted Constitution of the United States. Pennsylvania in particular had been a hard-fought battle, with its final delegate count being 46 for and 23 against. The next day, January 2, came the fourth state, Georgia. January 9 came Connecticut, and in February Massachusetts.

Maryland and South Carolina and New Hampshire all ratified within the coming months, bringing the count to nine of thirteen, each with their own suggested revisions or addendums.

THE MONTH OF JUNE SAW A BITTER FIGHT IN RICHMOND, VIRGINIA, FOR ITS very existence. The 170-strong delegates met from June 2 to June 27, with a final vote for ratification being evenly split, 89 aye to 79 nay. One the leaders of the nays was Patrick Henry, that prickly firebrand who hadn't lost his spark even a decade later.

It was a grueling convention, with each day dedicated to the problems of each individual article. Debates stemming from the phrase "We the People" in the very preamble brought shouting matches as Henry declared it was not appropriate for some Philadelphia convention to say they are "the People." He called the entire process "absurd" and of "lunatic" proportions, to have it ratified and *then* amended. "Were I about to give away the meanest particle of my own property, I should act with more prudence and discretion," he declared on Monday, June 9. "My anxiety and fears are great lest America, by the adoption of this system, should be cast into a fathomless bottom."[36]

And so it went for a month, with Henry and sizable opposition on one end, and James Madison and sizable support on the other, until June 25, 1788, when a vote for ratification barely passed.

Immediately after, a committee of twenty was appointed to address proposed amendments. In total, a "bill of rights" with twenty items was proposed, along with twenty amendments to the Constitution. "That there are certain natural rights, of which men," said the first, "when they form a social compact, cannot deprive or divest their posterity; among which are the enjoyment of life and liberty, with the means of acquiring, possessing, and protecting property, and pursuing and obtaining happiness and safety." The second declared "all power is naturally vested in and consequently derived from the people." Refusal to quarter soldiers, freedom of religion, a ban on excess bail, the right of peaceful assembly, freedom of speech, all proposed here in these "essential and unalienable rights of the people."[37]

Virginia became the tenth state to say aye to the Constitution.

AS SHE NEARED THE AGE OF EIGHTY, HER HEALTH DETERIORATING, MARY Ball Washington knew her time of judgment was soon to arrive. That decay of nature that Sir Matthew Hale wrote of long ago was finally catching up with her.

It was time to set her final stage of life in motion.

Comprising fewer than six hundred words, Mary Washington's last will and testament was drafted by her next-door neighbor and friend James Mercer. No relation to her physician Hugh Mercer, James lived at the St. James' House on Charles Street, a typical and small Colonial home built two decades earlier.

Mary began her last will and testament with an exhortation appropriate for a religious widow: "In the name of God! Amen! I, Mary Washington of Fredericksburg in the County of Spotsylvania, being in good health . . ."

Good health or not, her will, signed and sealed on May 20, 1788, would be executed before too much longer.[38]

THE URGENCY OF HIS MOTHER'S ILLNESS PLAGUED GEORGE, WHO WAS IN Mount Vernon tending to his fields on the date that Mary's will was drafted. Presumably receiving word she was writing her will, he decided to pay her a visit, one of the very few these last years.

On the morning of Tuesday, June 10, George and Martha traveled south to Fredericksburg, spending the night at Colonel Thomas Blackburn's in Prince William County. The next morning, they went out again, hoping to catch William Fitzhugh for dinner across the Rappahannock, but due to his absence, they instead crossed the river and "alighted at my Mothers and sent the Carriage & horses to my Sister Lewis's—where we dined and lodged."[39] He, of course, as was his duty, gave her some money, this time 4 pounds, 4 shillings.[40]

Legend has it that during this time—specifically during the

ministry of Thomas Thornton, which started in January 1788—when George visited his aging mother, he attended church the following Sunday. Here, his diary entry supported it. And here, Philip Slaughter, an early historian of Saint George's Church in Fredericksburg, published a rather fun anecdote, by way of future judge John Tayloe Lomax, who was seven years old at the time. "On Sunday," Slaughter said, "in accordance with the uniform habit, he attended the Episcopal Church, and so great was the crowd drawn together by his presence, that some of the timbers in the gallery which had not been well adjusted, settled into their places with a slight crush, which excited great alarm for fear the church was falling." The congregation quickly evacuated through any door and window they could, until a clerk calmed them down and asked that they return.[41]

The trip down and back put George behind on his correspondences; he offered apologies in letters to Henry Knox, Richard Henderson, and James Madison for not responding sooner to their missives. In all replies, he noted his visit to his mother (to Madison he wrote, "My aged and infirm mother").[42]

AS THE YEAR 1788 CAME TO A CLOSE, SO TOO DID THE DEBATE ON RATIFICATION: on July 26, New York became the eleventh state to accept the substance of the Constitution with requested alterations. Through the following months, until September 13, there waged a congressional debate about the next step in the formation of the United States.[43]

In particular, that would be the election of its first president.

Chapter 13

Unto Dust Shalt Thou Return

THE ILLNESS, DEATH, AND WILL OF

MARY BALL WASHINGTON

1788–1789

"She is translated to a happier place."

The year 1789 began with a political revolution: on January 7, George Washington was chosen to lead the newly formed country. "All 69 electors voted for Washington, making him the only president in American history to win unanimously," wrote Ron Chernow.[1] (Only two presidents have come close since then, Richard Nixon in 1972 and Ronald Reagan in 1984, both winning all but one state.)

John Adams of Massachusetts, the runner-up with 34 electoral votes (but carrying no majority in the states), was elected as the nation's first vice president.

"NATIVE SON ELECTED FIRST PRESIDENT!" shouted the *Virginia Herald and Fredericksburg Advertiser* in big, bold type. Indeed, the Fredericksburg community was in joyous celebration that their very own, though expected, had become the leader of a new nation.[2]

One cannot underestimate how obvious the choice was. From the moment the Constitution was ratified, it was clear Washington would be the man to lead. Everybody knew his résumé. Washington of course privately saw it with trepidation: "In answer to the observations you make on the probability of my election to the

Presidency (knowing me as you do)," he wrote to the Marquis de Lafayette a year earlier, "I need only say, that it has no enticing charms, and no fascinating allurements for me."[3] Months later, to Alexander Hamilton, he wrote, "I should unfeignedly rejoice, in case the Electors, by giving their votes in favor of some other person, would save me from the dreaded Dilemma of being forced to accept or refuse."[4] In the *Federalist Papers*, making the case for a Constitution to replace the Articles of Confederation, Hamilton, along with John Jay and James Madison, had always imagined Washington when describing the Chief Executive.

In subsequent months, a fierce debate erupted on how to address the president himself. John Adams and others suggested "His Highness the President of the United States and Protector of Their Liberties." Others, like Thomas Jefferson, wanted "His Excellency," or "His Honor," or "Esquire." Still others wanted a simple "George W., President of the U.S.," with no title. It was eventually decided by Washington himself a simple and straightforward solicitation: "Mr. President."[5]

Washington understood the weight of the responsibility that would befall him on April 30, 1789, Inauguration Day. In response to an earlier letter from Senator John Langdon of New Hampshire, he officially accepted the results of the election: "Having concluded to obey the important & flattering call of my Country, and having been impressed with an idea of the expediency of my being with Congress at as early a period as possible," he said simply, "I propose to commence my journey on thursday [*sic*] morning which will be the day after to morrow."[6]

However, before he could leave for New York, to end his retirement for yet another time, he would need to wrap things up at home, his beloved Mount Vernon, and make the journey to see his mother for what he probably knew would be the last time.

HE AGAIN PLACED THE EVER-DIFFICULT MOUNT VERNON PLANTATION IN the hands of George Augustine Washington, realizing he would not be able to return for some years from New York City. Just several months earlier, in December, he saw his potato crop fail, as it had continuously lost money. Letters went out to George Augustine about the estate and how to pay for the failing crops amid a poor economy.

On Saturday, March 7, 1789, he left for Fredericksburg "to discharge the last Act of personal duty, I may, (from her age) ever have it in my power to pay my Mother."[7]

He arrived in the evening of that same day. The *Virginia Herald and Fredericksburg Advertiser* painted the scene of his visit with jubilance. "His Excellency GEORGE WASHINGTON arrived in town from Mount Vernon, and early on Monday morning he let out on his return. The object of his Excellency's visit was, probably, to take leave of his aged mother, sister, and friends."[8]

What a momentous occasion, it implied, to have the favorite son, the newly elected leader of America, return home.

In reality, there was probably little joy when he saw his ailing mother.

The cancer that they had discovered several years earlier would have taken a toll at the age of eighty. Cancer, a terrible disease that baffled doctors—how to prevent or treat it was unknown.

With no medication or remedies identified at the time to relieve her of pain and discomfort, she would most likely have been thin, frail, and easily fatigued.

When mother and son saw each other for the last time, he sat with her, informing her of the grandest responsibility and honor that had been bestowed on him. George Washington Parke Custis, who would have been at the tender age of nearly eight at the time, relayed the story. "The people, madam," son said to mother, "have been pleased, with the most flattering unanimity, to elect me to the chief magistracy of these United States."

Did she know what a chief magistrate was, in the context of the new nation? Perhaps she thought her son had been appointed king. If her supposed Loyalist views were true, the thought of King George Washington, the First of America, may have sounded quite pleasing.

Either way, she understood its significance and the faith that the electors, the people, had placed in her son. Whether king or president or simple "leader," to rule a country was a grand duty. "But before I can assume the functions of my office," George continued, "I have come to bid you an affectionate farewell. So soon as the weight of public business, which must necessarily attend the outset of a new government, can be disposed of, I shall hasten to Virginia, and—"

"—and you will see me no more," Mary interrupted. The reason for the trip here was clear and weighed heavily. George was saying a final goodbye.

Mary continued, "My great age, and the disease which is fast approaching my vitals, warn me that I shall not be long in this world. I trust in God that I may be somewhat prepared for a better."

It is not hard to imagine that Mary reached out and placed her hands on George's gray head, in the ancient gesture of benediction. "But go, George, fulfill the high destinies which Heaven appears to have intended you for; go, my son, and may that Heaven's and a mother's blessing be with you always," she said.

Here, Custis added a little of his own commentary, imagining son and mother embracing, the man's head resting gently on her shoulder. "That look which could have awed a Roman senate in its Fabrician day, was bent in filial tenderness," he said.[9]

Further lore, via a descendant, provided an apocryphal but closing tale for the final encounter of mother and son. As George got up and headed to the door, with Mary by his side, he reached into his purse and brought out some gold coins. It had been second nature, really, to give her money every time he visited, and this was no exception. But again, apocryphally, this time, she refused. "I don't need it, my son," she said. "My wants are few, and I think I have enough."

George responded, "Let me be the judge of that, mother, but whether you think you need it or not, keep it for my sake."[10]

He recorded in his ledger his giving of 8 pounds, 8 shillings, to his mother.[11] Whether or not the conversation took place, he still felt it was, as a dutiful son, his responsibility to provide her even in her dying days. He left for Mount Vernon within two days—and George never saw Mary again.

SEVERAL DAYS LATER, ROBERT LEWIS, BETTY'S SON, WAS OFFERED THE POSItion of secretary to his uncle the president, a large responsibility that Robert realized he was "under a thousand obligations" to thank him for. Mary got word of young Robert's moving northward to New York with Martha and George, and offered him her carriage. "My grandmother was very well disposed to lend the carriage," he wrote, "but on condition that it should be returned when of no further use to my aunt."[12] The offer of the carriage was never accepted, however, for some unknown reason.

IN THE INTERVENING MONTHS BETWEEN HER FINAL GOODBYE TO HER SON and her death, the cancer would have become more and more aggressive, to the fear of her, her children, and her doctors. Mary knew the outcome, as newspapers announced many times the deaths of others by that terrible disease.

Cancer had been of note by this time. In 1738, an unnamed woman in Maryland noticed a small tumor on her breast and chalked it up to a wound. Some months later, as it was reported, "the whole Breast became Tumeified, and to such a Bulk, that there remained no Hope to repel contained Matter." But the disease was determined incurable.[13]

Through decades of research and trial and error, there were reports of cures discovered all around the colonies and in England. One woman from Prince Edward County, Virginia, thought she had discovered a cure, publicly advertising it. However, it was reported soon after that it "has been proved ineffectual, and so far from producing a perfect cure." The cure as Constant Woodson attested only hurt the patient. James Kirk, a relative of the patient, posted a warning to anyone who had or wanted help.[14]

MARY MAY HAVE FIRST DISCOVERED THE LUMP IN HER BREAST SOMETIME in the previous few years. Women's clothing was designed with a loose front, so it was easy to notice any peculiarities or abnormalities. Mary may have just thought it a mole or something that would go away in time.

She had hoped, and perhaps prayed, that it wouldn't be what she had feared. That lump did not go away, but instead spread.

AS HEROIC AND GODLIKE AS EARLY BIOGRAPHIES DESCRIBE HER, MARY'S death was not one of an Amazon warrior or of some grand Homeric heroine. She died in pain. Probably a great deal of pain.

The spring of 1789 passed, and the cancer would have spread through her body. Dr. Charles Mortimer and Dr. Elisha Hall, both Fredericksburg physicians and friends of the Washingtons, tried unsuccessfully to slow it if not cure it. All sorts of medicines were tried and failed. Dr. Hall even solicited help from Dr. Benjamin Rush, his cousin and friend, in Philadelphia. Rush responded in a letter on July 6, 1789: "My dear Kinsman," he addressed it respectfully.

The respectable age and character of your venerable patient lead me to regret that it is not in my power to suggest a remedy for the cure of the disorder you have described in her breast. I know nothing of the root that you mentioned is found in Carolina and Georgia, but from a variety of inquires and experiments I am disposed to believe that there does not exist in the vegetable kingdom an antidote to cancers. All the supposed remedies I have heard of are compounds of some mineral caustics. . . . From your account of Mrs. Washington's breast, I am afraid no great good can be expected from the use of [Dr. Martin's powder]. Perhaps it may cleanse it and thereby retard its spreading. You may try it, diluted in water. Continue the application of opium and camphor, and wash it frequently with a decoction of red clover. Give anodynes when necessary, and support the system with bark and wine. Under this treatment, she may live comfortably many years and finally die of old age.[15]

While mastectomies did occur, in the age before anesthesia, with the advance of both the cancer and the patient's age, it was decided not to put her under the knife. Surgery was highly dangerous, and to operate on Mary would in and of itself be something no doctor would have wanted to do. Dr. Hall, hearing the advice of his cousin, probably then applied a root of vegetable and arsenic with all other suggestions. "She may live comfortably many years," Rush had said.

She did not. She died within two months.

As her son, the president, was becoming familiar with his new job in New York, as Congress was passing legislation on taxes and tariffs and beginning the establishment of departments (the first, the Department of State, was established on July 27), family and neighbors and doctors rushed to the aid of Mary, trying all sort of experiments. George himself fell extremely ill, with a tumor growing on his left leg and an accompanying fever. A New York City physician, Dr. Samuel Bard, diagnosed the president with anthrax—a disease with symptoms of swelling and painful lumps.[16]

As the United States was only a couple months into its brand-new experiment, Mary lay dying. Betty was by her side. She wrote to her older brother, the president of the United States, on July 24, "I am sorry to inform you My Mothers Breast still Continues bad. God only knows how it will end; I dread the Consequence. she is sensible of it & is Perfectly resign'd—wishes for nothing more than to keep it Easy." And in what George may have thought was a surprise, his mother cared little about herself, but about him and his illness: "She wishes to here [*sic*] from you, she will not believe you are well till she has it from under your Hand." These were her only known last words, to care about her son's well-being and her son's health, knowing her own death was imminent.

The doctors had wanted hemlock to ease the pain, Betty noted, but there was none available.[17]

What reply Washington gave in response to the letter, if he gave one, is lost.

THE FINAL MONTH OF MARY'S LIFE WAS ONE OF INCREASING DETERIORA-tion. She saw her end coming, and there was no cure that any doctor could offer. She was suffering, in terrible pain.

As a woman of faith, she would have looked to God in these final weeks. She would have meditated with tried and true passages from the Bible or from Sir Matthew Hale's own work. Though death was the end of her earthly life, it opened up another doorway for her eternal soul.

Good Christians of the New World believed Christ was the pathway to eternal life. Of all tenets of Christianity, this was the most plain. Mary, who had been faithful to her religion her whole life, would have known this. "I am the resurrection, and the life: he that believeth in me, though he were dead, yet shall he live," Christ said to Martha, sister of Lazarus. As Mary, who lay on her deathbed, understood, as Lazarus himself was resurrected, she too would be at the end of days, with all the righteous, in the new kingdom of heaven.

About fifteen days before her death, around August 10, 1789, her health took a step toward her final hours. Writing to President George Washington, Burgess Ball, a distant relative of Mary's and the husband of her grandchild Frances Washington, relayed her deterioration: "For abt 15 days she has been deprived of her speech."[18]

Reverend Thomas Thornton of Saint George's Episcopal Church would have received word of her impending end. Though there is no formal last rite or sacrament in the Protestant or Anglican Church as there is in the Catholic Church, the Episcopalian Book of Common Prayer had several sections for the communion and preparation for death. Either Reverend Thornton or another minister according to her faith would have recited the blessings and her communion, as was the tradition.[19]

WITH HER FAITH IN GOD SECURE, AROUND AUGUST 20, MARY BALL WASHING-ton fell into a coma. The speech impediment and coma (as Burgess Ball noted to George Washington, "she has remain'd in a Sleep") may have been signs of a stroke, only decreasing her time on earth. Within five days, at around three o'clock in the afternoon on August 25, she took her last breath and finally passed away, over the age of eighty. Her exact age is still not known. Her suffering of these final years had finally ended, and her loved ones believed she entered into eternal life with the God that she worshiped all her life.

From an unknown date around 1708 to August 25, 1789, Mary Washington, née Ball, lived a long and active life. An orphan, a wife, a mother, a widow, an owner of plantations, manager of servants and slaves, she was at times overbearing to her children, at times protector of them. She had been a widow for over forty years, refusing to remarry and looking toward her children's future.

"She has lived a good Age &, I hope, is gone to a happier place than we live at present in," wrote Burgess Ball. As one of her final acts, Mary had asked Burgess to write her son of her impending end. It was a moment of unusual care in her later life, to let her son know of her health. "Mrs Lewis being in much trouble, and all her Sons absent, she requested I wd write to you on the Subject."

Enclosed with the letter was a copy of the last will and testament for Mary Ball Washington.[20]

HER SHORT WILL WAS THE FINAL AFFIRMATION TO HOW SHE SAW HER RE-maining surviving children. As George was the eldest, and perhaps her favorite, still, after all these decades, she gave him much of her remaining estate.

"I give to my son General George Washington," it said, "all my land in Accokeek Run, n the County of Stafford, and also my negro boy George, to him and his heirs forever. Also my best bed, bedstead, and Virginia cloth curtains (the same that stands in my best bedroom), my quilted blue and white quilt and my best dressing glass."

While this did not seem to be a great deal, she had very little in property at the end, only confined to her Fredericksburg house. To Charles in western Virginia, she simply gave her slave Tom. To Betty, she gave her carriage and horse. To her daughter-in-law Hannah, she gave a purple cloak; to Corbin Washington, her grandson and son of John Augustine, she gave her "negro wench, old Bet," a riding chair, and two horses. Other slaves were given to other grandchildren, no more than one per, and various glassware and beds and china were also bestowed. Mary equally divided her "wearing apparel" among her granddaughters as they saw fit, though she allowed Betty, if she should "fancy any one two or three articles," to claim some as her own.

Finally, at the end, she nominated her son, calling him again, "General George Washington," to be the executor of her will.[21] Of course, at the time the will was written, George was in northern Virginia. Mary's bequest to give her remaining land and best items to him, was much.

MARY'S AGE WAS UNKNOWN. FROM THE EARLIEST INFORMATION AVAILABLE, it was presumed to be somewhere between eighty and eighty-five. George Washington Parke Custis said she was eighty-five, a grand age.[22] James Thomas Flexner wrote eighty-one years old at her passing,[23] which may have been accurate if she were born in 1708 before late August. The *Gazette of the United States*, on September 9, listed her as eighty-two, adding a short epitaph to her obituary:

> *O may kind heaven, propitious to our fate;*
> *Extend THAT HERO'S to her lengthen'd date;*
> *Through the long period, healthy, active, sage;*
> *Nor know the sad infirmities of age.*[24]

The *Virginia Herald*, two days after her death, published her obituary, a lengthy praise of her equally lengthy life, and gave her age as eighty-two years old. (So did Washington Irving.[25]) The paper called her "the venerable mother of the illustrious President of the United States, after a long and painful indisposition, which she bore with uncommon patience." It continued,

Tho' the pious tear of duty, affection and esteem, is due to the memory of so revered a character, yet our grief must be greatly alleviated from the consideration that she is relieved from all the pitiable infirmities attendant on an extreme old age—it is usual when virtuous and conspicuous persons quit this abode, to publish an elaborate panegyric on their character—suffice it to say, she conducted herself through this transitory life with virtue, prudence and Christianity, worthy the mother of the greatest hero that ever adorned the annuals of history.[26]

For a woman kept at length by her son, her obituary certainly appreciated her presence on earth.

ON SEPTEMBER 1 PRESIDENT WASHINGTON WAS IN NEW YORK CITY ENJOYing dinner with two friends, Baron Friedrich Wilhelm von Steuben, a military genius who helped form the Continental Army, and General Arthur St. Clair. During the evening's activities, a courier arrived with the letter from Burgess. Washington's nephew and secretary, Betty's twenty-year-old son Robert Lewis, recorded in his diary that it was an "event so long expecd [it] could not create so much uneasiness as [a] person less advanced in life."[27] In an undated letter to his mother, Robert wrote that "my uncle immediately retired to his room, and remained there for some time alone."[28]

GEORGE WASHINGTON WAS PRESIDENT OF THE UNITED STATES, AND EVERYday responsibilities of that honor did not cease. Three days after receiving the news, he appointed Andrew Ellicott to survey western New York State for a sum of $1,125. The business of government went forward, and it could not stop.[29]

SOON AFTER RECEIVING THE NEWS, HE BOUGHT, FOR 5 POUNDS, 13 SHILlings, and 10 pence, mourning cockades, sword knots, and black armbands. He ordered all men in the presidential household in New York City to wear the cockades on their arm or hat, and for women to wear black necklaces. Further, formal receptions were canceled for the following three weeks.[30]

Before the order, however, to suspend receptions, there was a very somber one on Tuesday, September 8, which, as reported in the *Gazette of the United States*, was attended by several congressmen "and other respectable persons." It was held "in American mourning. The silent mark of respect, flowing spontaneously from the hearts of freemen sympathizing with him in this domestic misfortune, manifests sentiments and emotions which no language can express in a manner so unequivocal and delicate."[31] It was as public an expression of mourning as George Washington had for his mother, Mary Ball Washington, whose body was many hundreds of miles south. It also contradicted other historians' slants. One historian waved away any mourning from George beyond the expected, believing he was instead indifferent to it all. "Washington spoke no eulogies, told no fond anecdotes of the hard, sometimes shrewish mother who had served as the lifelong whetstone of his anger," he wrote flatly.[32]

On September 13, a week after the reception, the now motherless president finally sent a letter to Betty. It was a lengthy and heartfelt letter between brother and sister. "My dear Sister," he began,

Colonel Ball's letter gave me the first account of my Mother's death. . . . Awful, and affecting as the death of a Parent is, there is consolation in knowing that Heaven has spared ours to an age, beyond which few attain, and favored her with the full enjoyment of her mental faculties, and as much bodily strength as usually falls to the lot of four score. Under these considerations and a hope that she is translated to a happier place, it is the duty of her relatives to yield due submission to the decrees of the Creator—When I was last at Fredericksburg, I took a final leave of my Mother, never expecting to see her more.

With that sad letter written, Washington went on to more earthly matters, that of her will. What his mother left him was, frankly, embarrassing to him; he was confused to why she chose him, the most well-off of her children, to receive the best of her property. George plainly stated his role as executor was "impossible." If an appraisal could be finalized without him, then the will could be finished and the heirs receive what was given. The slaves that were not named had to work until the harvest's end, and any debt that Mary had, of which she said she owed little or none, could be paid off with crops, horses, or other means.

"She has had a great deal of money from me at times, as can be made appear by my books," he added in frustration. "In short to the best of my recollection I have never in my life received a copper from the estate—and have paid many hundred pounds (first and last) to her in cash." He continued, a little calmed down now, to get it off his chest, "However I want no retribution— I conceived it to be a duty whenever she asked for money, and I had it." Further, the Fredericksburg house would preferably be sold and not rented, giving final closure to a tired son of that stress.

Finally, George asked Betty, with the help of Charles Carter (husband of Betty's daughter, another Betty, and the son of the legendary Robert "King" Carter) and Colonel Burgess Ball, to help speed the process as sufficiently as possible. And even though Mary was gone, her troubles would follow him for the remainder of the year.[33]

GEORGE WASHINGTON HAD NOT YET LEARNED OF MARY'S DEATH BY THE time she was buried. Held on or about August 28, the funeral for

Mary Washington was relatively extravagant for a woman of her age, while certainly equal to a woman of her worth. Fredericksburg's business was suspended for that day. Joseph Berry, the town crier, rang the bell of Saint George's Episcopal Church twice, then again once for each year she had lived. He had done the same the day she died, and was paid over 1 pound, 15 shillings for "carrying messages and Tolling the Bell."[34] Her coffin, made by neighbor and furniture maker James Allen, cost 8 pounds, 6 shillings, and 4 pence—not particularly cheap, suggesting something more ornate than an average coffin.[35]

Her casket, at the close of the ceremony, was buried, as she requested, near her favorite spot: Meditation Rock, where she prayed many hours during the Revolution.

THE ACTUAL GRAVESITE OF MARY BALL WASHINGTON IS LOST TO HISTORY. She was buried somewhere near Kenmore plantation, near her Meditation Rock and near her later constructed monument, unknown specifically where. The site where her body lay was lost soon after her death, to the surprise of many of Washington's historians. "The grave of Washington's mother is marked by no visible object, not even a mound of earth, nor is the precise spot of its locality known," wrote a shocked Jared Sparks in 1827, a mere forty years after her death. "The burial ground is on the western border of the town, and was formerly on the estate belonging to the Lewis family. . . . For a long time a single cider-tree was the only guide to the place; near this tree tradition has fixed the grave of Washington's mother, but there is no stone to point out the place, nor any inequalities in the surface of the ground."[36]

Michelle Hamilton of the Mary Washington House in Fredericksburg has "a couple hunches" of where she might be, but without ample archaeological digging, no one can find it. "I would love to see that happen," said Hamilton, excitedly. "I have dreams of the American version of Richard III," where excavators uncovered the medieval English king's remains under a parking lot.[37] Through this discovery, they learned about his DNA, his illnesses, and his facial structure. The same could be done for Mary: how far along her breast cancer was, what her face looked like. Even her height, unknown, could be determined. Did she have weak bones by the end of her life? All this, and so much more, could be figured out, closing so much mystery around her, humanizing her beyond the words on pages in books.

THE *VIRGINIA HERALD AND FREDERICKSBURG ADVERTISER*, ON OCTO-ber 15, 1789, announced that there, "on Thursday, the 29th instant, will be sold at the plantation, about 4 miles below this town, late the property of Mrs. Mary Washington, deceased, ALL the stocks of Horses, cattle, sheep and hogs, plantation utensils of every kind, carts, hay and fodder. That trouble of collections &c. may be avoided, they will be sold to the highest bidder, for ready money. All persons having claims against the deceased are requested to bring them in properly attested to."[38] The ad was placed by Betty and showed an amassed amount of property that may have surprised the average native of Fredericksburg. In fact, her will had left out a number of items, including a basket, two dinner tables, four old rugs, and much more. There was so much that a memorandum was made to list and catalog it all, and for how much it was sold, if recorded. There was one mattress sold to Burgess Ball, two walnut chairs, one carpet sold to Dr. Elisha Hall at 45 shillings; to Lawrence Lewis, Betty's son, went a wheel for 3 shillings, a tea chest for about 9, a spice box for 18.[39]

Months had gone by, and months followed. Mary's body had been buried, in the ground of the town she lived for nearly all her life. But her ghost remained active for the family.

Two weeks before the advertisement ran, Betty reviewed the finances that Mary had left behind. She was disheartened. "The Docters [sic] bills is more than I expectted [sic]," she wrote to President Washington. She had owed both Dr. Hall and Dr. Charles Mortimer ("Mortemores," she called him) 45 and 22 pounds, respectively. "The Debts I think will be upwards of one Hundred Pound." Burgess Ball, Betty, George, and Charles Carter decided at that point to run the advertisement in the *Virginia Herald* in hopes of paying off the debt. Any expectations that the crops would cover this final headache were crushed when a nasty frost hit the region and destroyed half the tobacco and much of the corn while still in the ground. There was, however, some good news: Mary's two lots in Fredericksburg were valued at 450 pounds, a testament to the value of the property so close to town.[40]

President Washington, recovering from an abscess unrelated to the influenza epidemic that had hit the northern states, read the letters from Betty and from Ball and Carter, and realized he was still needed as executor. As much assistance as the two men were offering his sister, there were still legal hurdles that needed his attention.

Two months passed, and George celebrated his first Christmas as president, visiting St. Paul's Chapel in New York City. In the months following Mary's death, the bureaucratic framework of the federal government increased.

AT TIMES, THE WORRY OF HIS MOTHER'S ESTATE MAY HAVE BEEN THE FUR-thest thing from his mind. George Washington had a country to run, and this work as president took priority over the task of being a will's executor. Due to the duties of the country, he did not reply for nearly a month to an overdue update from Burgess Ball the day after Christmas. "The Negroes are at length divided," he wrote. "We concluded to divide the Negroes into four parts by wch it [means] you have a Fellow call'd Dundee," along with a slave named George. The slave George had been Mary's legally, but had been living at Mount Vernon since the 1770s as rented property. He had married and fathered three children in the decades following, and Master George agreed to keep the slave George and his family. Burgess admitted, happily, that "none of the Families are parted," except for the spouses or children not legally part of Mary's estate.

On the Fredericksburg home, Ball noted that the initial valuation of 450 pounds in two years' time was too high by anyone else's count, and that no one would bite unless it was lower. Charles Carter, however, was interested, and compromised (to which George agreed) at 350 pounds in three years. "I do not know any Person who will give so much," Ball said.[41]

By March 1790, with months going by and other duties preventing this increasingly long chapter from ending, George Washington and Charles Carter agreed to the amount for the lots, and a deed was processed in the spring of 1790.

THE DEED MAY HAVE BEEN OUT OF HIS HANDS, BUT AGAIN, HIS MOTHER'S ghost still haunted him. For as luck would have it, within the three years, Charles Carter had not completed payment. "The credit you gave me on the purchase of yr property in this place, expired last spring," Carter wrote apologetically in October 1793. "It is a subject of great regret to me that any delay should have taken place in the fullfilment of my contract, tho' I assure you that circumstances quite unforeseen by me, have occasion'd it." He continued, saying that he was to sell the property to a Richard Dobson of Cumberland County.[42]

As he had so well in life, George Washington was still looking after the affairs of his mother.

By this point, Mary had been dead over four years. George had been reelected to the presidency. The Bill of Rights was completely ratified. Rhode Island and Vermont, the fourteenth state, had ratified the Constitution. Kentucky became the fifteenth state, taken from western Virginia. The United States dollar was standardized throughout the country. In France, the monarchy was overthrown and the Reign of Terror under the First Republic was well under way. Benjamin Harrison had died in 1791, and John Hancock died only a week before Carter's letter. The capital of the United States had moved from New York City to Philadelphia.

The country created moved on, and so did George. In a letter on May 29, 1794, he wrote Carter. It was a short letter, clearly wanting to put the whole matter behind him. Among the short passages he wrote, he forgave Carter of the remainder of his debt. He waved it away.[43]

Finally, his mother's posthumous problems were behind him, and he could serve the rest of his term as president. When he finished that, he could finally, again, for a final time, retire in peace as Mary rested in peace.

But Mary's legacy was yet to come.

Epilogue:
A Monumental Legacy

" 'Mary, the Mother of Washington.'
No eulogy could be higher."

───────────

H onored Madam." That was how George Washington always greeted his mother in his letters. It was a phrase that revealed *her* austerity and *his* awe of the woman who had been called "commanding" and "strict" throughout her life, as well as "truly kind." Through all of this contradictory information, however, an image emerges of a woman whose fierceness and individuality are obvious antecedents to the same qualities in her bold son.

While, in George, these qualities evinced themselves through nobility and sacrificial leadership, Mary's authority was used quite differently. Her small-time tyranny hung over the son who would go on to found the freest nation on Earth. He was the man who, when he could have seized kingship for himself, walked away. His stubbornness and singularity that such independence required must have come from the single mother who ran a plantation on her own.

THE FIRSTBORN SON OF MARY BALL WASHINGTON, GEORGE WASHINGTON, former president, former commander of the Continental Army, died at the age of sixty-seven on December 14, 1799, at his Mount Vernon home. A mere three months before, he had lost his younger brother, Charles, who died at age sixty-one. His only surviving sister, Betty, sixty-three years old, had died two years earlier. "The death of near relations, always produces awful, and affecting emotions, under whatsoever circumstances it may happen," George wrote to Burgess Ball after learning of Charles's death.[1] The oldest child, the son of Mary and Augustine, was the last to die.

Washington was left unburied for three days, as he had requested, to make certain he was indeed dead. In accordance with his will, he was interred at Mount Vernon plantation, ensuring he would never again leave that dearly loved home. Throughout the years, he had left his home and his beloved Martha so many times; he was determined to stay for good. His remains are there today, as are those of Martha, and other family members, including his cousin Bushrod, and his brother John Augustine, to name a few. Etched into the stone above his crypt: "Within this Enclosure Rest the remains of Gen'l George Washington."

WHEN HE DIED, THERE WERE 317 SLAVES ON HIS FIVE FARMS, SOME OF WHOM were too old or too young to work. Of those 317 slaves, he freed them all upon his death. Martha closed the master bedroom door at Mount Vernon where he had died. She began sleeping in an attic bedroom. She would live another two and a half years, and die at the age of seventy.

AS FOR HIS BOYHOOD HOME, FERRY FARM PLANTATION HAD PASSED FROM history, and from the Washingtons, all the way back in 1774, when Dr. Hugh Mercer, physician, neighbor in Fredericksburg, fellow Fredericksburg Masonic member, and friend to George, bought it all for 2,000 pounds, to be paid in five annual installments. Washington had trouble selling the farm, it seemed, evidence of its failing status; Mercer's offer came two years after the first advertisement in the *Virginia Gazette*. Within the week of the agreement, George wrote to him: "The Land may be conveyed to you at any time, & for this purpose I will bring down my title papers & leave them with you."[2]

Washington went to Fredericksburg on May 13, at which time

he presumably gave his papers to Mercer, finalizing all documents during the short stay.

And on May 14, it was no longer Washington's. Mercer wanted but never managed to move there, and hoped to build a settlement on the farm.

It was all for naught.

Then general Hugh Mercer was mortally wounded in the Battle of Princeton, New Jersey, on January 3, 1777. The farm passed on to his heirs, and an executor of Mercer's will, George Weedon, agreed to pay the remaining debt.[3]

Ferry Farm through the rest of history experienced decline and war. The Mercers never lived on the property itself, and it was sold to Winter Bray in 1846. The Bray family kept the land until the 1870s. Strangely enough, the farm itself never saw action during the Revolutionary War. Only during the American Civil War a century later, when Union troops stationed themselves by Ferry Lane near the Rappahannock, was George Washington's boyhood home occupied by a military presence. The Rappahannock was a necessary waterway for the Union Army to secure. Canisters, unspent bullets, and buttons were later found buried in the farm's soil.[4]

IMMEDIATELY UPON HIS DEATH, HE WAS ELEVATED TO A NEAR GODLIKE STATUS in the new nation, both figuratively and literally. David Edwin's engraving from 1800 showed a burial-shrouded Washington, crowned with laurel from angels, as he ascended into heaven from Mount Vernon. The more famous depiction of this veneration is the not-so-subtly named *The Apotheosis of Washington* by Roman-born Constantino Brumidi, painted in 1865, which reigns overhead in the Capitol rotunda. Brumidi had previously worked under Pope Gregory XVI in the Papal States, so his imagery of Washington, ascending on a cloud in royal purple surrounded by the goddesses Victoria and Libertas, had heavy religious overtones. Horatio Greenough's larger-than-life sculpture, now sitting in the Smithsonian, celebrated the centennial of Washington's birth, and depicted him more as a Julius Caesar or Zeus than a president, bare chested with a shroud draped over one arm raised in blessing, a sheathed sword in another.

A lock of Washington's hair was found recently in an eighteenth-century almanac, given by the president while he was still alive, and it made worldwide news.[5] It was and has been kept very much like a relic, coveted.

The phrase "George Washington slept here" entered the common vernacular many years ago, in and around the trail from Mount Vernon to Fredericksburg to Williamsburg, and New York, New Jersey, Massachusetts, and Connecticut, as well. The claim is often true, as plaques on houses and hearths proudly announce. Washington was also known to pace in these houses, late into the evening. Many have records of the leader of his country walking the floors, heard by residents, still as part of their lore. Hollywood once made a comedy entitled *George Washington Slept Here* based on a play of the same name. His friend, former Revolutionary War major general, member of the Continental Congress, and father of Robert E. Lee, Henry "Light Horse" Lee, was commissioned by the Sixth Congress to write a eulogy of the first president of the United States. In his elegant and touching tribute, he famously wrote that Washington was "first in war, first in peace, and first in the hearts of his countrymen."

GEORGE WASHINGTON WAS PART OF, AND WITNESS TO, THE FORMATION OF the United States of America. He was the grand commander and grand president of the newly formed nation and the savior of Americans' very freedoms. As Americanism took hold, Manifest Destiny and westward expansion, immigration and the desire for American freedom grew; so, too, did his status. "The first name of America, not only is, but always will be, that of Washington," wrote John Frederick Schroeder in 1850, in his preface to his compilation of Washington's sayings. "We pronounce it with filial reverence, as well as gratitude."[6]

And yet Mary, his mother, floated at the time above the truth, almost mystical, hagiographical. Fanciful books from Benson Lossing or Sara Pryor in the mid- to late 1800s certainly contributed to the early legends of her status as *the* mother of Washington—almost a goddess in her own right. But these works, popular and definitive for over a century, were oft ignored once twentieth-century histories, including Douglas Southall Freeman's and James Thomas Flexner's multivolume and momentous biographies, rewrote the depiction of Mary from matron to shrew.

Where were the monuments? Where were the cities in her name? Unverified and unlikely lore said that her son, the president, wished for a monument to be erected soon after her death, and some even reported that Congress passed a resolution. None exist.[7]

George Washington Parke Custis, that great-grandson who knew and remembered her fondly from his childhood, was the first to lead the hagiographical charge. His remembrance of his great-grandmother, partly published in 1821 in the *National Gazette*, brought Mary and her grave to national attention again. "Had she been of the olden time," he had written, "statues would have been erected to her memory in the capitol [*sic*], and she would have been called the Mother of Romans." He lamented the lack of recognition that she deserved. "When another century shall have elapsed, and our descendants shall have learned the true value of liberty, how will the fame of the paternal chief be cherished in story and in song, nor will be forgotten her. . . . Then, and not till then, will youth and age, maid and matron, aye, and bearded men, will pilgrim step, repair to the now neglected grave."[8]

Years passed, and the grave continued to go unmarked. To this day, no one knows exactly where Mary was buried, and no remains have ever been found. The supposed neglect of the area was exaggerated sometimes, as by an anonymous writer in the *Richmond Visitor and Telegraph*, who wrote, "Her grave is in a deserted, dreary, solitary field—the mound of earth that was originally raised over her sacred remains is now washed away."

Fredericksburg's own newspaper, *Political Arena*, chided this writer: "He surely never saw the spot . . . or his description would have been different." Instead of a solitary field, it was on a "beautiful knoll" within the town, "in a highly cultivated and fertile field." The paper continued, definitively saying the grave "stands in no danger of being profaned."[9]

By 1830, the people of Fredericksburg wanted something to commemorate her. Over a century later, James Thomas Flexner wrote a damningly negative view of the delay: "She had, by her complaints that had made her seem a Tory, so disassociated herself from her son's charisma that it was not until she had been long dead, and her living presence had been completely eroded away by the Washington legend, that any marker was placed on her grave."[10] There may be a grain of truth in this. People have long memories. But though forty years had passed since her death by then, and thirty had passed since her eldest son's death, there were still people who probably grew up knowing her or were raised by those who knew her—and they wanted some commemoration. Hardly a lack of "her living presence" was felt in the town. The desire was there. The hope was there. They simply needed the money for to erect something befitting the Mother of Washington.

It came when Silas Burrows of New York, a man of considerable wealth, wrote to the Fredericksburg mayor, wishing "to erect a monument over the remains and to rescue from oblivion the sacred spot where reposes the great American mother, Mary, the mother of Washington."[11]

Within two years, a cornerstone had been placed, laid by President Andrew "Old Hickory" Jackson in a special ceremony on May 7, 1833. Secretary of War Lewis Cass, Attorney General Roger Taney, and private secretary Andrew Jackson Donelson also attended. The ceremony was followed by a barbecue and procession of monument committee members, clergy, architects, mayor and councilmen, teachers and students, musicians, and others.

The president spoke with vigor and strength, a long eulogy for both Mary and her son. "In the grave before us, lie the remains of his Mother. Long has it been unmarked by any monumental tablet, but not unhonored. You have undertaken the pious duty of erecting a column to her name, and of inscribing upon it, the simple but affecting words, 'Mary, the Mother of Washington.' No eulogy could be higher, and it appeals to the heart of every American."

ON JUNE 7, 1924, PRESIDENT CALVIN COOLIDGE SIGNED INTO LAW A BILL TO create a replica of the house where Washington was born at Wakefield in Westmoreland County, Virginia, but it was later criticized as being a "totally inaccurate reproduction of the birth house, both in design and location."[12] It was opened, however, in July of 1931, just months before the anniversary of Washington's two hundredth birthday. While excavating, they did discover priceless artifacts in the area, including glass, tableware, and other household accessories possibly used by the Washington family themselves.[13]

WHAT OF HER CHARACTER? SHE DID MORE THAN GIVE BIRTH TO AND RAISE George. She shaped him. "Tradition says, that the character of Washington was strengthened, if not formed, by the care and precepts of his mother. She was remarkable for the vigor of her intellect and the firmness of her resolution," said President Jackson. There were still people who remembered her from their own childhoods. President Jackson understood this. "It is impossible to avoid the conviction," he said to his audience, that based on what people had told him, her "principles and conduct . . . were closely interwoven with the destiny of her son. . . . Look back at the life and conduct of his mother, and at her domestic government . . . and they will be found admirably adapted to form and develop, the elements of such a character."

With the laying of the cornerstone, he finished his speech, saying, "Fellow-citizens, at your request, and in your name, I now deposit this plate in the spot destined for it; and when the American pilgrim shall, in after ages, come up to this high and holy place, and lay his hand upon this sacred column, may he recall the virtues of her, who sleeps beneath, and depart with his affections purified, and his piety strengthened, while he invokes blessings upon the Mother of Washington."[14]

ANOTHER FOUR YEARS PASSED, AND THE MONUMENT, A MAUSOLEUM IN shape with Doric columns, was partly built. And construction . . . stopped. Suddenly and without a clear explanation. Many, of course, have been offered, from Silas Burrows's wealth suddenly falling, to being rejected in his courtship of a young girl descended from Mary, to the bank actually losing Burrow's donation, to the death of the monument's contractor, Rufus Hill. Whatever actually happened, by 1840, the monument was left incomplete, with an estimated fifty-foot-long pillar lying on the ground.

By 1848, construction of the Washington Monument in the District of Columbia—a giant obelisk worthy of the giant founder—began. It would take four decades for its completion, however, and when the construction restarted, the builders could not match the stone precisely. Mary's remained untouched and unfinished.

Decades followed and the monument was still left incomplete, with nothing or no one to account for it. It was an embarrassment to many of Mary's early biographers to see it become nothing more than rubble. The 1860s saw the Civil War tear the country apart, and Fredericksburg was no exception to what occurred; cannon and rifle fire riddled the stationary stone in cross fire. Nearby Kenmore, Betty and Fielding's house, was used as a field hospital and damaged horribly by the ghastly war that pitted brother against brother.

THE OVERSIGHT WAS RECTIFIED BY THE 1890S. IT WAS APPROPRIATE THAT renewed interest occurred the century after her death with almost lightning speed. The National Mary Washington Memorial Association was formally organized on February 22, 1890, in honor of George Washington's birthday. It was an organization six months in the making, of women who, according to the *Washington Post* at the time, "organized themselves into an association for the purpose of erecting a monument to Washington's mother, and maintaining and preserving the same in good order."[15]

It couldn't have come at a better time, as an ad in the *Post* in early 1889 placed the grave and land up for public auction. One woman, Margaret Hetzel, enclosed a dollar and a plea to the *Post* to start a fund to save the land. She became the first donor to what would become, in the years following, a successful organization (along with the contemporary and new Daughters of the American Revolution). Many prominent women joined, including the author of the oft-sourced and hagiographical *Story of Mary Washington*, Mary Terhune. It was a cross-country call, the first by women to honor a woman, in recognition of Mary Ball Washington.

After thoughtful design and approval by all, with a clear mirroring of the Washington Monument, a new cornerstone was placed on her grave on October 21, 1893. It was a much less elaborate celebration compared to sixty years earlier, but it was still held in reverence by the people of Fredericksburg. A copy of *The Story of Mary Washington*, old pennies, a membership card to the Mary Washington association, pictures of the old house, and many other historical items dear to Mary or the Fredericksburg residents were placed at the granite base, as mementos or relics of the woman they all honored.[16]

The white obelisk was completed, piercing the sky, at four o'clock on December 22, 1893. Four days later, the protective box covering the shaft was removed.

Finally, over one hundred years after her passing, the monument of Mary Washington was completed. It was simply inscribed, "Mary, the Mother of Washington."

A PUBLIC DEDICATION OCCURRED ON MAY 10, 1894, IN THE PRESENCE OF PRESident Grover Cleveland and members of the cabinet and Chief Justice Melville Fuller. "The town is rapidly filling up," reported the *Lisbon Sun*, as the ceremony was to prove popular.[17] Luncheon was served at Mary's old house, as a procession worthy of her brought the entire town together in joyous celebration. President Cleveland, echoing his predecessor, spoke to a large audience, saying he was there "to perpetuate the memory of her who gave birth to the leader of the American armies in the mighty struggle; fashioned his genius, moulded his character, formed his soul for good, and inspired him for the work of liberating his people from the fetters of tyranny."

He added, "Here under this bright sky and in these clear sunbeams the first monument is to be dedicated in remembrance of this noble American matron; . . . [built] by a glorious band of women who determined to rescue the memory of the mother of Washington from the corroding hand of time."[18]

President Cleveland continued for some time, emphasizing the importance of Mary to her son, and how her son transformed not just the nation that he led, but the entire world. Right after, Mann Page, the grand master of the Grand Lodge of Virginia, officially dedicated the monument.[19]

It was complete. The monument stands today in blazing white, surrounded by a wrought iron fence. It was the first national monument in the nation dedicated to a woman.

IN 1954, PRESIDENT DWIGHT EISENHOWER VISITED FREDERICKSBURG ON Mother's Day, May 9, after earlier commemorating the anniversary of the Richmond Light Infantry Blues farther south. He signed the register of, but did not visit, her house, then spoke at her gravesite in a special ceremony. It was a short but sweet visit on a rainy day, only about twenty minutes long but noticeably important. Mary, he said to a crowd of several thousand, "gave us the greatest American of them all, and on this day we think of his great attributes of patience and of his courage in adversity, his perseverance, his faith in him, his God, and his country. If we remember those things, ceremonies like this will take on added significance." Wearing a white carnation pinned on his black suit, he placed a wreath at the foot of her monument.[20]

Fredericksburg and the area around Lancaster has honored the mother of the first president in other ways. In 1938, the State Normal and Industrial School for Women at Fredericksburg, established in 1908, was renamed to Mary Washington College (it was renamed University of Mary Washington in 2004). In Lancaster, a road runs several miles through beautiful countryside, called, simply, Mary Ball Road. In 1943 was completed the SS *Mary Ball*, a Liberty cargo ship, owned and operated by the United Fruit Company for three years before being placed on the National Defense Reserve Fleet (she was scrapped in 1972). A small but dedicated volunteer group runs the Mary Ball Washington Museum and Library in Lancaster. Ball family descendants live today in and around the Northern Neck and Middle Peninsula of Virginia.

And her house in Fredericksburg, which her son George had purchased for her, is and has been a popular tourist attraction in a town steeped in Revolutionary War and Civil War history.

THAT IS THE LEGACY SHE LEFT. BUT IT'S MUCH MORE THAN THAT TOO. Roads and museums and monuments and a college are in her name. The way historians viewed her has changed more than a little since her son's first biographer, Parson Weems, wrote his book. Who made this Founding Father, but his mother, the woman who influenced him for good or for ill? "Mrs. Mary Washington," wrote historian and author Washington Irving, "is represented as a woman of strong plain sense, strict integrity, and an inflexible spirit of command."[21] Positive or negative in tone, that has remained a constant in her histories. Whether Freeman or Pryor, both say she was a strong character who forged her son's destiny.

Yet for all her fierceness, Mary was ultimately indebted to and entrenched in the establishment of her youth. She did not take those vital next steps into a broader world, as George did. A member of the establishment religion, she raised George Washington in faith, giving him those first lessons about his place in society. But when the time came for George to follow his calling into peril and rebellion, Mary sought to check him. In a way, George's first battle for freedom was from his own mother. For all Mary's faith and devotion, in the end her motherly love may have been as much about authority as affection.

It was thus that George Washington came to manhood, under the "Maternal hand," as he wrote, of Mary Ball Washington. It made him the man he was: stubborn, singular, awe-inspiring, and *independent.*

Mary Ball Washington, the Mother of the Father—born and raised and lived and died in Virginia, with no travel overseas or out of the colony, then commonwealth, living to at least eighty, a widow, the Rose of Epping Forest, the Belle of the Northern Neck—lives on in the hearts and minds of her fellow Americans.

She was and always will be George Washington's Honored Madam.

Acknowledgments

Sometimes I think writing a book is vaguely akin to making a documentary, but with a lot more research and drafting and many, many more hours. There are many behind-the-scenes players who work tirelessly and with so much expertise; without their help, the project would remain just a dream. They help in the making of this book not for themselves, but for the facts. As you don't see the Sturm und Drang that goes into making a documentary, it is the same with a book. But as Winston Churchill said, of writing a book, eventually "you fling him out to the public." Before any flinging is done, however, there are some people I need to thank.

Thank you to my family, especially my wife, my best friend, the center of my universe, Zorine. I never want to think what my life would be like without my wonderful life partner. She edits, questions, cajoles, and makes better all my works; but on this one, Zorine really dug in and helped create the final book you are now reading.

And to our children, Matthew, Andrew, Taylor, and Mitchell. Thanks for all the support and love. Thank you also to my mother, Barbara Shirley Eckert, who has always been there for me and who did so much preliminary research on Mary Ball Washington. Thank you, Mother. All have helped me tremendously in this project, supporting me from the beginning.

Thank you, Scott Mauer, my researcher, who has now become a very dear friend. Scott is a superb individual in every measure of a man and a professional. We've now worked on three books together and in each case, Scott has come through like a champ. If I was to undertake any task whatsoever, one of the first people I would call would be Scott.

A special comment must be made here about Hannah Long, my editor at HarperCollins. Hannah worked tirelessly, patiently, calmly, professionally, and cheerfully to see this book though from beginning to end. She simply made this book better. Thank you, Hannah.

Also thank you to Doug Bradburn, president of Mount Vernon, who was so helpful in every request. Thank you, Doug.

Thank you, Bridget Matzie, a great friend and a great agent. None of this would have happened without Bridget. Thank you. Thank you to my partner Patricia Halgas and the staff at Shirley & Banister, including Kevin McVicker, Cassandre Durocher, Sarah Selip, Carolyn Lisa, Taylor Shirley, Bryson Boettger, and Francesca Goerg, for your patience and support and encouragement.

Thank you to Gay Hart Gaines, the Distinguished Visiting Lecturer in American History of the Mount Vernon Ladies' Association, formerly president of that illustrious association, who is a dear friend and close confidant. A strong and smart woman in her own right. Thank you to Jon Meacham for being a pal and supporter. Jon is one of America's best historians. Each time we talk, I learn something new.

Stewart McLaurin, of the White House Historical Association, has been a friend and supporter for many years. John Heubusch, Joanne Drake, Melissa Giller, and Fred Ryan, all of the Ronald Reagan Presidential Library, were all delighted to hear of this project, because they love history as I do. Thank you, all. And thanks to Maureen Mackey, who also edited this book. This is the second book of mine that she has worked on and, like last time, she would not accept anything less than excellence.

Michelle Hamilton, manager of the Mary Washington House, was eager to help and was thrilled to hear of this project. She offered unpublished anecdotes, wisdom, and documents which she had collected for this book. As manager for several years, she has placed an indelible mark on this historical house in Fredericksburg. Thank you for everything, and for your enthusiasm in answering all questions.

Thank you to Genevieve Bugay, manager of the Hugh Mercer Apothecary Shop, who gave insight into eighteenth-century medicine and provided much-needed material on the health of Mary Ball Washington, especially as her cancer affected her through her later life.

Thank you to Katie King, manager, and Daniel Hawkins, guide, at the Rising Sun Tavern, Charles Washington's former home in Fredericksburg. The tavern was converted after Charles Washington had moved to Charles Town, West Virginia, and the tour and knowledge learned of social customs were absolutely necessary to paint a living picture. Thank you, both.

The Mary Washington House, Hugh Mercer Apothecary Shop, the Rising Sun Tavern, and St. James's House are owned and operated by the Washington Heritage Museums, based in Fredericksburg. Thank you, all, especially executive director Anne Darron.

Thank you to Gary Hayes, guide at Kenmore, and all the managers at that historical home, for providing great insight to the life of both Fielding and Betty Lewis. Those at Ferry Farm too, who honor Washington.

Thank you, all others at Mount Vernon, including historian Mary Thompson and director of archaeology Luke Pecoraro. Also to Angelica Yost and Samantha Snyder of Mount Vernon. The Mount Vernon Ladies' Association has been working for over 160 years promoting and helping the nation learn of its first president. Curt Viebranz, former president of Mount Vernon, has been an enthusiastic supporter of my works. Thank you, Curt.

Both Karen Hart and Barbara Whitbeck at the Mary Ball Washington Library and Museum in Lancaster were jovial and knowledgeable in their history of George Washington's mother. The Library offered many documents—some published, some not—all conveniently ready to view. Their hard work helped enormously. Thank you.

Thank you, Jack Bales, Reference and Humanities Librarian, and Gary Stanton, professor, at University of Mary Washington, who helped guide this book with sourced documents and period knowledge.

Gary introduced the project to J. Travis Walker, archivist at the Fredericksburg Circuit Court Archive, who had previously worked as a researcher for Ferry Farm, George Washington's childhood home and where Mary lived for many decades of her life. And thanks to Carolyn Parsons, Head Archivist, University of Mary Washington.

Thank you to Hilary Derby, a local historian in Northumberland County who helped track down seventeenth-century documents at the Northumberland Public Library, and who was eager as any to dive right in.

Thank you, Mike Moses, secretary at Fredericksburg's Masonic Lodge, of which George Washington and so many of his compatriots were members. Thank you, Shelby L. Handler.

Thank you to Michele Lee Silverman, research services librarian at the Society of the Cincinnati in Washington, D.C.

Thank you to Melissa Corben, supervisor at the George Washington Birthplace National Monument.

Nancy Moore, Virginiana Room Manager at the Central Rappahannock Regional Library, and Judy Chaimson at the Central Rappahannock Heritage Center both provided unpublished documents of the *Virginia Herald* and other rare items from and after the period. Thank you.

Thank you to Kate Collins at the David M. Rubenstein Library at Duke University and thank you, Kathryn Blizzard of the Washington Papers at the University of Virginia, for hunting down primary documents scattered throughout the country.

Sophie Lillington, manager of the Epping Forest Museum in London, England, answered questions of the Balls' and Washingtons' English ancestry. Thank you, Sophie, for giving a distinctly "English" vibe.

Thank you to Graham T. Dozier and Jameson Davis at the Virginia Museum of History & Culture (paintings).

Thank you, of course, to Eric Nelson and Eric Meyers, my managing editors at HarperCollins, who have been patient and supportive of me every step of the way. And thank you to Adam Bellow, who supported this project from the beginning.

Some other supportive family and friends need mentioning. They include Michael McShane, George F. Will, Newt Gingrich, Kevin Kabanuk, John Morris, Patricia Gallagher, Rick and Sue Johnson, Mike Murtagh, Roshan and Perin Bhappu, Tom and Lyn Finnigan, Frank Donatelli, Carl Cannon, Paul Bedard, Jennifer Harper, Dan Wilson, and Charles Pratt.

My deepest appreciation to all.

I am known as a Reagan biographer, having written four books on our fortieth president with several more to come. But I have also written other books as well, on Newt Gingrich, on World War II, and other topics. Each has been immeasurably pleasurable. So it is also with *Mary Ball Washington*. Mary Ball Washington is a compelling though underreported and fascinating figure of history.

I hope with this book to have helped—in a small way—to rescue Mary from drifting into the forgotten sands of time.

Craig Shirley
Ben Lomond, California
2019

Notes

ABBREVIATIONS

The Diaries of George Washington—DGW

The Papers of Benjamin Franklin—PBF

The Papers of George Washington. Colonial Series—PGWCol

The Papers of George Washington. Revolutionary War Series—PGWRev

The Papers of George Washington. Confederation Series—PGWCon

The Papers of George Washington. Presidential Series—PGWPres

The Papers of George Washington. Retirement Series—PGWRet

The Papers of James Madison—PJM

The Papers of Thomas Jefferson—PTJ

The Papers of Thomas Jefferson, Retirement Series—PTJRet

WORKS OF INTEREST

George Washington by Henry Cabot Lodge (American Statesmen series; Boston: Houghton, Mifflin, 1889).

George Washington by William H. Rideing (True Stories of Great Americans series; New York: Macmillan, 1916).

Basic Writings of George Washington edited by Saxe Commins (New York: Random House, 1948).

The Story-Life of Young George Washington by Wayne Whipple (Philadelphia: John C. Winston, 1911).

PROLOGUE

1. Letter from Marquis de Chastellux to George Washington, August 23, 1789; George Washington Papers, Manuscript Division, Library of Congress, Washington, DC.

2. Patricia Brady, *Martha Washington: An American Life* (New York: Penguin, 2005), 69–70.

3. Jessie Biele and Michael K. Bohn, *Mount Vernon Revisited* (Charleston, SC: Arcadia, 2014), 11.

4. David McCullough, *1776* (New York: Simon & Schuster, 2005), 44.

5. David A. Clary, *George Washington's First War: His Early Military Adventures* (New York: Simon & Schuster, 2011), 9.

6. Clary, 8.

7. Dorothy Twohig, "The Making of George Washington," in *George Washington and the Virginia Backcountry*, ed. Warren R. Hofstra (Madison, WI: Madison House, 1998), 7.

8. Silas Weir Mitchell, *The Youth of Washington, Told in the Form of an Autobiography* (New York: Century, 1910), 76–77.

9. Letter from George Washington to Joseph Jones, February 11, 1783, manuscript/mixed material, Library of Congress, Washington, DC.

10. Letter from George Washington to Bryan Fairfax, March 1, 1778, *PGWRev*, vol. 14, ed. David R. Hoth (Charlottesville: University of Virginia Press, 2004), 9–11.

11. Bonnie Angelo, *First Mothers: The Women Who Shaped the Presidents* (New York: HarperCollins, 2000), 406.

12. Letter from George Washington to Mary Ball Washington, February 15, 1787, *PGWCon*, ed. W. W. Abbot, vol. 5 (Charlottesville: University of Virginia Press, 1997), 33–37.

13. Douglas Southall Freeman, *George Washington*, vol. 5 (New York: Charles Scribner's Sons, 1951), 491.

14. "The Story of Mary Washington," *American Monthly Magazine* (Washington, DC), vol. 2, January–June 1893, 735.

15. Margaret C. Conkling, *Memoirs of the Mother and Wife of Washington* (Auburn, NY: Derby, Miller, and Co., 1850), 20.

16. Freeman, *George Washington*, vol. 1 (New York: Charles Scribner's Sons, 1948), 193.

17. George Washington Parke Custis, *Recollections and Private Memoirs of Washington* (New York: Derby & Jackson, 1860), 131.

18. James Thomas Flexner, *George Washington: The Forge of Experience, 1732–1775* (Boston: Little, Brown, 1965), 41.

19. James Thomas Flexner, *George Washington and the New Nation (1783–1793)* (New York: Little, Brown and Company, 1969), 227.

20. Lydia H. Sigourney, *Great and Good Women: Biographies for Girls* (Edinburgh: William P. Nimmo, 1866), 124.

21. Custis, *Recollections and Private Memoirs of Washington*, 131.

22. Letter from George Washington to the Citizens of Fredericksburg, February 14, 1784, *PGWCon*, vol. 1, 122–23.

CHAPTER 1: VIRGIN LAND AND VIRGIN LOVE

1. James Horn, *Adapting to a New World: English Society in the Seventeenth-Century Chesapeake* (Chapel Hill: University of North Carolina Press, 1994), 159–77.

2. Edward G. Lengel, *First Entrepreneur* (Boston: Da Capo Press, 2016), 10.

3. Marion Harland, *The Story of Mary Washington* (Boston: Houghton, Mifflin, 1893), 3–4.

4. "New Historic Preservation Easements," *Notes on Virginia*, no. 32 (1989), 5–8.

5. F. L. Brockett and George W. Rock, *A Concise History of the City of Alexandria, Va.* (Alexandria: Gazette Book and Job Office, 1883), 6.

6. David Brown, "Skeleton of Teenage Girl Confirms Cannibalism at Jamestown Colony," *Washington Post*, May 1, 2013.

7. *Virginia Gazette*, March 10, 1774, May 16, 1755, June 31, 1771, April 15, 1773, April 1, 1780, April 21, 1768, Colonial Williamsburg Archives, Williamsburg, Virginia.

8. Joseph J. Ellis, *The Quartet: Orchestrating the Second American Revolution, 1783–1789* (New York: Alfred A. Knopf, 2015), 40.

9. Brockett and Rock, *A Concise History of the City of Alexandria, Va.*, 5–6.

10. Daniel Blake Smith, *Inside the Great House: Planter Family Life in Eighteenth-Century Chesapeake Society* (Ithaca, New York: Cornell University Press, 1980), 80.

11. Abigail Adams, "Letter from Abigail Adams to John Adams, 31 March–5 April 1776," Adams Family Papers: An Electronic Archive, accessed March 12, 2019, https://www.masshist.org/digitaladams/archive/doc?id=L17760331aa.

12. E. James Ferguson, *The American Revolution: A General History, 1763–1790* (Homewood, IL: Dorsey Press, 1974), 25.

13. Guy Stevens Callender, *The Economic History of the United States, 1765–1860* (Boston: Ginn, 1909), 70.

14. Ferguson, *The American Revolution*, 25.

15. Ferguson, 28.

16. John Smith, *Travels and Works of Captain John Smith*, ed. Edward Arber, vol. 1 (New York: Burt Franklin, 1910), lxviii.

17. Thomas Jefferson, *Notes on the State of Virginia* (Boston: Lilly and Wait, 1832), 77.

18. As quoted in Ferguson, *The American Revolution*, 47.

19. Ferguson, 51.

20. Sara Agnes Rice Pryor, *The Mother of Washington and Her Times* (New York: Macmillan, 1903), 50.

21. Dorothy Denneen Volo and James M. Volo, *Daily Life During the American Revolution* (Westport, CT: Greenwood Press, 2003), 287–88.

22. Ferguson, *The American Revolution*, 51.

23. Ferguson, 48.

24. Volo and Volo, *Daily Life During the American Revolution*, 289–91.

25. Ferguson, *The American Revolution*, 40.

26. *Virginia Gazette*, September 14, 1769, Colonial Williamsburg Archives.

27. *Virginia Gazette*, September 19, 1751, Colonial Williamsburg Archives.

28. Horn, *Adapting to a New World*, 149–50.

29. Diary entries, December 3, 1709, June 17, 1710, "William Byrd's Diary," *Africans in America*, PBS, https://www.pbs.org/wgbh/aia/part1/1h283t.html, excerpted from *The Secret Diary of William Byrd of Westover, 1709–1712*, Louis B. Wright and Marion Tinling, eds. (Richmond, VA: Dietz Press, 1941).

30. Ferguson, *The American Revolution*, 43.

31. Ron Chernow, *Washington: A Life* (New York: Penguin Press, 2011), 622–23.

32. Pryor, *The Mother of Washington and Her Times*, 158–59.

33. Ferguson, *The American Revolution*, 46.

34. Ferguson, 47.

35. Mary V. Thompson, *"In the Hands of a Good Providence": Religion in the Life of George Washington* (Charlottesville: University of Virginia Press, 2008), 16.

36. Thompson, *"In the Hands of a Good Providence,"* 17; Ferguson, *The American Revolution*, 56–57.

37. Thompson, *"In the Hands of a Good Providence,"* 2–3.

38. Rhys Isaac, *Worlds of Experience: Communities in Colonial Virginia* (Williamsburg, VA: Colonial Williamsburg Foundation, 1987), 33.

39. Thompson, *"In the Hands of a Good Providence,"* 15.

40. Thompson, 16–18.

41. John Tracy Ellis, *American Catholicism* (Chicago: University of Chicago Press, 1969), 21–25.

42. William Whitmore, ed., *The Colonial Laws of Massachusetts,* facsimile of the 1672 edition (Boston: City Council of Boston, 1889), 55.

43. "In and around Lancaster" (1981), unpublished manuscript, Mary Ball Washington Museum and Library, Lancaster, Virginia.

44. Pryor, *The Mother of Washington and Her Times,* 148–49.

45. *Virginia Gazette,* August 19, 1773, Colonial Williamsburg Archives.

CHAPTER 2: "TO LOOK TO THE SKY"

1. Charles Moore, *The Family Life of George Washington* (Boston: Houghton Mifflin, 1926), 22.

2. Conkling, *Memoirs of the Mother and Wife of Washington,* 15.

3. John Stotsenburg, "The Maternal Grandmother of George Washington," *Pennsylvania-German,* vol. 9, May 1908, 226.

4. Earl L. W. Heck, *Colonel William Ball of Virginia: The Great-Grandfather of Washington* (London: Sydney William Dutton, 1928), 15.

5. William Camden, *Remains concerning Britain* (London: John Russell Smith, 1870), 71; Charles W. Bardsley, *Curiosities of Puritan Nomenclature* (London: Chatto and Windus, 1880), 85.

6. Frances Smith, *Colonial Families of America* (New York: Frank Allaben Genealogical Company, 1909), 39.

7. Leonard Abram Bradley, *History of the Ball Family: Genealogy of the New Haven Branch* (New York: J. M. Andreini, 1916), 15.

8. Washington Irving, *Life of George Washington,* vol. 1 (New York: P. F. Collier & Son, 1856), 1.

9. John Frost, *Pictorial Life of George Washington* (Philadelphia: Leary & Getz, 1859), 14.

10. Bernard J. Cigrand, "Washington Not Real Name of Our First President," *New York Times,* February 19, 1911.

11. Frances Smith, *Colonial Families of America,* 40; Heck, *Colonel William Ball of Virginia,* 16.

12. As quoted in Wayne Whipple, *The Story-Life of Washington* (Philadelphia: John C. Winston, 1911), 4–7.

13. Heck, *Colonel William Ball of Virginia,* 16.

14. "Ball Family," Mount Vernon, http://www.mountvernon.org/digital-encyclopedia/article/ball-family/.

15. Mary Selden-Kennedy, *Seldens of Virginia and Allied Families,* vol. 1 (New York: Frank Allaben Genealogical Company, 1911), 150.

16. Heck, *Colonel William Ball of Virginia,* 27.

17. Heck, 27; Horace Edwin Hayden, *Virginia Genealogies* (Wilkes-Barre, PA: E. B Yordy, 1891), 51.

18. Hayden, *Virginia Genealogies,* 53.

19. Heck, *Colonel William Ball of Virginia*, 27.

20. Heck, 27–28.

21. George Cabell Greer, *Early Virginia Immigrants* (Richmond, Virginia: W. C. Hill Printing, 1912), 19.

22. George Washington Ball, *The Maternal Ancestry and Nearest of Kin of Washington: A Monograph* (Washington, DC: 1885), 6.

23. Horn, *Adapting to a New World*, 57–58.

24. John Evelyn, *Fumifugium: Or, The Inconveniencie of the Aer and Smoak of London Dissipated* (London: W. Godbid, 1661), 15–16, 23–24.

25. Heck, *Colonel William Ball of Virginia*, 28.

26. Heck, 29.

27. Heck, 29; Hayden, *Virginia Genealogies*, 51.

28. Lancaster County Court Records, order book 3, 16, 356–66, Mary Ball Washington Museum and Library, Lancaster, Virginia; Nina Tracy Mann, "Hannah Ball, a Colonial Matriarch," *Northern Neck of Virginia Historical Magazine*, vol. 22 (December 1972), 2317.

29. Horn, *Adapting to a New World*, 174–77.

30. Horn, 174–85.

31. Hayden, *Virginia Genealogies*, 51–52.

32. William Waller Hening, ed., *The Statutes at Large*, vol. 2 (New York: R. & W. & G. Bartow, 1823), 329–30.

33. Lancaster County Court Orders 1, 132, Mary Ball Washington Museum and Library, Lancaster, Virginia.

34. Lancaster County Court Orders 2, 22, 63; Court Orders 3, 76, Mary Ball Washington Museum and Library, Lancaster, Virginia.

35. Nina Tracy Mann, "William Ball, Merchant," *Northern Neck of Virginia Historical Magazine*, vol. 23 (December 1973), 2523.

36. Mann, "William Ball, Merchant," 2527.

37. Hayden, *Virginia Genealogies*, 51.

38. Heck, *Colonel William Ball of Virginia*, 32.

39. As quoted in Horn, *Adapting to a New World*, 288.

40. Lancaster County Will Book, vol. 5, 70–71; Northumberland Public Library, Heathsville, Virginia; Hayden, *Virginia Genealogies*, 50–51.

41. Mann, "Hannah Ball, a Colonial Matriarch," 2320.

42. Letter from Joseph Ball II to Joseph Chinn, May 13, 1755, in Correspondence of Joseph Ball, 1743–1780, 138, Manuscript Division, Library of Congress, Washington, DC.

43. Letter from Joseph Ball II to Joseph Chinn, October 22, 1756, 172, Manuscript Division, Library of Congress, Washington, DC.

44. Hayden, *Virginia Genealogies*, 56.

45. Freeman, *George Washington*, vol. 1, 531; Selden-Kennedy, *Seldens of Virginia and Allied Families*, vol. 1, 157–58.

46. Letter and attachment from H. Irvine Keyser II to Mr. Newton, November 7, 1966; The Society of the Cincinnati Archives, Washington, DC.

47. Heck, *Colonel William Ball of Virginia*, 35–36.

48. Moncure Daniel Conway, ed., *George Washington and Mount Vernon* (Brooklyn: Long Island Historical Society, 1889), xxiv.

49. Margaret Lester Hill, ed., *Ball Families of Virginia's Northern Neck: An Outline* (Lancaster: Mary Ball Washington Museum and Library, 1990), 145.

50. Hayden, *Virginia Genealogies*, 57–58.

51. Pryor, *The Mother of Washington and Her Times*, 26.

52. Hayden, *Virginia Genealogies*, 57; Conway, *George Washington and Mount Vernon*, xlv.

53. Flexner, *George Washington: The Forge of Experience*, 11.

54. Hayden, *Virginia Genealogies*, 57.

55. William Ball Wright, ed., *Ball Family Records: Genealogical Memoirs of Some Ball Families of Great Britain, Ireland, and America*, 2nd ed. (York: Yorkshire Printing Company, 1908), 183–84.

56. Edward North Buxton, *Epping Forest* (London: Edward Stanford, 1905), 4–7.

57. Harland, *The Story of Mary Washington*, 8.

58. Albert Welles, *The Pedigree and History of the Washington Family* (New York: Society Library, 1879), iii.

59. Welles, *The Pedigree and History of the Washington Family*, 7.

60. Ethel Armes, *The Washington Manor House: England's Gift to the World* (New York: The American Branch of the Sulgrave Institution, 1922), 8.

61. H. Clifford Smith, *Sulgrave Manor and the Washingtons* (London: J. Cape, 1933), 170–71.

62. G. Douglas Wardrop, "English Home of the Washingtons Dedicated to Peace," *New York Times*, January 4, 1914, 3.

63. "George Washington, Descendent of the Saints," *Catholic World*, April–September 1916, 140–41.

64. Woodrow Wilson, *George Washington* (New York: Harper & Brothers, 1897), 15.

65. Frank E. Grizzard Jr., *George Washington: A Biographical Companion* (Santa Barbara, CA: ABC-CLIO, 2002), 331; Peter A. Lillback with Jerry Newcombe, *George Washington's Sacred Fire* (King of Prussia, PA: Providence Forum Press, 2006), 83.

66. Willard Sterne Randall, *George Washington: A Life* (New York: Henry Holt, 1997), 10.

67. Moore, *The Family Life of George Washington*, 17.

68. Armes, *The Washington Manor House*, 18.

69. Henry Cabot Lodge, *George Washington*, vol. 1 (American Statesmen series; Boston: Houghton, Mifflin, 1899), 36.

70. Moore, *The Family Life of George Washington*, 19.

71. Frank W. Hutchins, "George Washington's Forefather Aided in the Founding of Maryland," *Sunday Star* (Washington, DC), October 28, 1934, 13.

72. Charles E. Hatch Jr., *Chapters in the History of Popes Creek Plantation*. Division of History, Office of Archeology and Historic Preservation, National Park Service (Washington, DC: National Park Service, 1968), 5–9.

73. Harland, *The Story of Mary Washington*, 29.

74. Hening, *The Statutes at Large*, 330–31.

75. Paul Leicester Ford, *The True George Washington* (Philadelphia: J. B. Lippincott, 1911), 16; Abby Sage Richardson, *The History of Our Country* (Boston: H. O. Houghton, 1875), 135–36.

76. Pryor, *The Mother of Washington and Her Times*, 16–17.

77. Joseph Dillaway Sawyer, *Washington* (New York: Macmillan, 1927), 53.

78. Joseph Meredith Toner, ed. *Wills of the American Ancestors of General George Washington* (Boston: New-England Historic Genealogical Society, 1891), 3–5; Hatch Jr., *Chapters in the History of Popes Creek Plantation*, 15–16.

79. Moore, *The Family Life of George Washington*, 20.

80. Toner, *Wills of the American Ancestors of General George Washington*, 4.

81. Toner, 18–20, 22–23.

82. Toner, 9–10.

CHAPTER 3: THE ROSE OF EPPING FOREST

1. Horace Edwin Hayden, "Mary Washington," *Magazine of American History*, vol. 30, nos. 1–2, July–August 1893, 50.

2. Nancy Byrd Turner, *The Mother of Washington* (New York: Dodd, Mead, 1930), 3.

3. Motley Booker and James F. Lewis, "Cox's Old Place Now Yeocomico View Farm in Cherry Point," *Bulletin of the Northumberland County Historical Society*, vol. 21, December 1984, 43.

4. Benson J. Lossing, *Mary and Martha: The Mother and the Wife of George Washington* (New York: Harper & Brothers, 1886), 7.

5. Harland, *The Story of Mary Washington*, 5.

6. Henry Dudley Teetor, "The Mother of Washington," *Spirit of '76*, March 1898, 1.

7. William Waller Hening, ed., *The Statutes at Large*, vol. 1 (Philadelphia: J. B. Lippincott, 1911), 439.

8. Horace Edwin Hayden, "The Maternal Grandmother of Washington," *Pennsylvania-German*, vol. 9, no. 7, July 1908, 315.

9. Hayden, "Mary Washington," 51–52.

10. Turner, *The Mother of Washington*, 6.

11. Turner, 7.

12. Alice Morse Earle, *Child Life in Colonial Days* (London: Macmillian, 1899), 20–21, 35.

13. Lossing, *Mary and Martha: The Mother and the Wife of George Washington*, 7–8.

14. Pryor, *The Mother of Washington and Her Times*, 32.

15. Harland, *The Story of Mary Washington*, 6.

16. As quoted in Hayden, *Virginia Genealogies*, 58–59.

17. Inventory of Joseph Ball's estate, July 25, 1711, Mary Ball Washington Museum and Library Archives, Lancaster, Virginia.

18. Paula S. Felder, *Fielding Lewis and the Washington Family: A Chronicle of 18th Century Fredericksburg* (Fredericksburg, VA: The American History Company, 1998), 12.

19. Felder, *Fielding Lewis and the Washington Family*, 13.

20. Pryor, *The Mother of Washington and Her Times*, 37–38.

21. Pryor, 33.

22. George William Beale, "An Unwritten Chapter in the Early Life of Mary Washington," *Virginia Magazine of History and Biography*, vol. 8, no. 3, January 1901, 284.

23. Pryor, *The Mother of Washington and Her Times*, 36.

24. Lossing, *Mary and Martha: The Mother and the Wife of George Washington*, 9.

25. Hening, *The Statutes at Large*, vol. 2, 517.

26. William Meade, *Old Churches, Ministers and Families of Virginia*, vol. 2 (Philadelphia: J. B. Lippincott & Co., 1900), 128.

27. Turner, *The Mother of Washington*, 35.

28. Earle, *Child Life in Colonial Days*, 64–66.

29. Pryor, *The Mother of Washington and Her Times*, 39–43.

30. Earle, *Child Life in Colonial Days*, 122.

31. Randall, *George Washington: A Life*, 16.

32. Conkling, *Memoirs of the Mother and Wife of Washington*, 16.

33. Ella Bassett Washington, "The Mother and Birthplace of Washington," *The Century*, vol. 43, April 1892, 830.

34. James Walter, *Memorials of Washington and of Mary, His Mother, and Martha, His Wife* (New York: Charles Scribner's Sons, 1887), 131–32.

35. Robert C. Auld, "Sir Joshua Reynolds," *Self Culture*, vol. 4, no. 3, December 1896, 245.

36. Felder, *Fielding Lewis and the Washington Family*, 14.

37. Pryor, *The Mother of Washington and Her Times*, 56–57.

38. Pryor, 62–63.

39. Virginia Carmichael, *Mary Ball Washington* (Richmond: Association for the Preservation of Virginia Antiquities, 1967), 11–13.

40. John Stotsenburg, "The Maternal Grandmother of George Washington," 227.

41. Flexner, *George Washington: The Forge of Experience*, 11.

42. Turner, *The Mother of Washington*, 79.

43. Randall, *George Washington: A Life*, 16.

44. Alexis de Tocqueville, *Democracy in America*, trans. Henry Reeve, vol. 2 (New York: J. & H. & G. Langley, 1840), 208–9.

CHAPTER 4: THE MARRIAGE OF MARY BALL AND AUGUSTINE WASHINGTON

1. Harland, *The Story of Mary Washington*, 156.

2. Lossing, *Mary and Martha: The Mother and the Wife of George Washington*, 67–68.

3. Portrait by Robert Edge Pine, c. 1786; Prints and Photographs Division, Library of Congress, Washington, DC.

4. Charles Henry Hart, "An Inquiry into the Authenticity of the Portrait of Mary Ball, the Mother of Washington," *New York Genealogical and Biographical Record*, vol. 49, no. 2, April 1918, 150–54.

5. Taylor Soja, "Mary Ball Washington," Mount Vernon website, archived at https://web.archive.org/web/20180511092026/https://www.mountvernon.org/library/digitalhistory/digital-encyclopedia/article/mary-ball-washington.

6. "Exhibits Are Extended through December 20," *Rappahannock Record*, December 11, 2014, B3.

7. Michelle Hamilton, in discussion with Scott Mauer, April 24, 2018.

8. Harland, *The Story of Mary Washington*, 157.

9. Letter from Nellie Parke Custis Lewis, March 16, 1851, Mary Washington House Archives, Fredericksburg, Virginia.

10. Harland, *The Story of Mary Washington*, 33.

11. "Mildred Gale (1671–1701)," *White Haven and Western Lakeland*, http://www.whitehavenandwesternlakeland.co.uk/people/mildredgale.htm.

12. Lillback and Newcombe, *George Washington's Sacred Fire*, 88–89.

13. Chernow, *Washington: A Life*, 5.

14. Hatch Jr., *Chapters in the History of Popes Creek Plantation*, 27.

15. Lengel, *First Entrepreneur*, 9.

16. Lillback and Newcombe, *George Washington's Sacred Fire*, 89; Hatch Jr., *Chapters in the History of Popes Creek Plantation*, 28–30.

17. Ford, *The True George Washington*, 17.

18. Chernow, *Washington: A Life*, 5.

19. Jared Sparks, *The Life of George Washington* (Auburn, NY: Derby & Miller, 1851), 3–4.

20. Westmoreland Orders, 1722–1731, 95; Library of Virginia, Richmond, Virginia.

21. Hatch Jr., *Chapters in the History of Popes Creek Plantation*, 54.

22. James M. Swank, *History of the Manufacture of Iron in All Ages* (Philadelphia: American Iron and Steel Association, 1892), 265.

23. Hatch Jr., *Chapters in the History of Popes Creek Plantation*, 80–81.

24. Moore, *The Family Life of George Washington*, 21–22.

25. Moncure Daniel Conway, *Barons of the Potomack and the Rappahannock* (New York: Grolier Club, 1892), 52–53.

26. Freeman, *George Washington*, vol. 1, 55.

27. "Ironworkers in American History," *Bulletin of the American Iron and Steel Association*, September 30, 1894, 284.

28. Charles Brown, "General Washington," *New England Historical and Genealogical Register*, January 1857, 4–5.

29. Washington, "The Mother and Birthplace of Washington," 832.

30. Turner, *The Mother of Washington*, 100.

31. Turner, 95.

32. Harland, *The Story of Mary Washington*, 37–38.

33. Turner, *The Mother of Washington*, 101–2.

34. Randall, *George Washington: A Life*, 16.

35. *Virginia Gazette*, November 19, 1772, Colonial Williamsburg Archives; Smith, *Inside the Great House*, 129.

36. Washington family Bible, Mount Vernon Archives, Mount Vernon, Virginia.

37. Turner, *The Mother of Washington*, 104.

38. H. Ragland Eubank, *The Authentic Guide Book of Historic Northern Neck of Virginia* (Richmond: Whittet & Shapperson, 1934), 64.

39. Smith, *Inside the Great House*, 155.

40. Eubank, *The Authentic Guide Book of Historic Northern Neck of Virginia*, 155.

41. Smith, *Inside the Great House*, 162.

42. Harland, *The Story of Mary Washington*, 45–46.

43. Custis, *Recollections and Private Memoirs of Washington*, 141.

44. Freeman, *George Washington*, vol. 1, 45.

45. Washington, "The Mother and Birthplace of Washington," 838.

46. Custis, *Recollections and Private Memoirs of Washington*, 126–28.

47. Moore, *The Family Life of George Washington*, 25.

48. Letter from Jean-Baptiste Donatien de Vimeur to George Washington, February 12, 1781, manuscripts/mixed material, Library of Congress, Washington, DC.

49. Diary entries, February 11, 1798, February 12, 1798, *DGW*, ed. Donald Jackson and Dorothy Twohig, vol. 6 (Charlottesville: University of Virginia Press, 1979), 282; letter from Henry Knox to George Washington, February 11, 1790; manuscripts/mixed material, Library of Congress, Washington, DC.

50. Turner, *The Mother of Washington*, 114.

51. Joshua Hempstead, *Diary of Joshua Hempstead of New London, Connecticut* (New London, Connecticut: New London County Historical Society, 1901), 57.

52. Smith, *Inside the Great House*, 29–30.

53. Smith, 28.

54. William Meade, *Old Churches, Ministers and Families of Virginia* (Philadelphia: J. B. Lippincott & Co., 1861), 170.

55. Irving, *Life of George Washington*, vol. 1, 16.

56. Benson J. Lossing, *The Home of Washington* (New York: W. A. Townsend, 1865), 34.

57. Elbridge S. Brooks, *The True Story of George Washington* (Boston: Lothrop Publishing, 1895), 16–17.

58. Pryor, *The Mother of Washington and Her Times*, 80–81.

59. J. Paul Hudson, *George Washington Birthplace: National Monument, Virginia* (Washington, DC: Government Printing Office, 1956), 13.

60. Frost, *Pictorial Life of George Washington*, 16.

61. Flexner, *George Washington: The Forge of Experience*, 12.

62. Smith, *Inside the Great House*, 28–29.

CHAPTER 5: IN THE SHADOW OF THE EMPIRE

1. Harland, *The Story of Mary Washington*, 36–37.

2. Brown, "General Washington," 1–4.

3. Mason Locke Weems, *The Life of George Washington* (Philadelphia: Joseph Allen, 1837), 7–8.

4. John Stevens Cabot Abbott, *George Washington: Or, Life in America One Hundred Years Ago* (New York: Dodd & Mead, 1875), 12–13.

5. David Humphreys, *Life of General Washington*, ed. Rosemarie Zagari (Athens: University of Georgia Press, 1991), 6–7.

6. Benjamin Franklin, *The Autobiography of Benjamin Franklin* (New York: Charles E. Merrill, 1892), 14.

7. Thomas Jefferson, *Autobiography of Thomas Jefferson* (New York: G. P. Putnam's Sons, 1914), 4–5.

8. Chernow, *Washington: A Life*, 6.

9. James M. Volo and Dorothy Denneen Volo, *Family Life in 17th- and 18th-Century America* (Westport, CT: Greenwood Press, 2006), 193.

10. Jacqueline Howard, "These Are the States Where Infant Mortality Is Highest," *CNN*, January 4, 2018.

11. Volo and Volo, *Family Life in 17th- and 18th-Century America*, 193; Judith Waizer Leavitt, *Brought to Bed: Childbearing in America, 1759–1950* (New York: Oxford University Press, 1986), 15–16.

12. Philippe Ariès, *Centuries of Childhood: A Social History of Family Life*, trans. Robert Baldick (New York: Alfred A. Knopf, 1962), 128.

13. John Locke, *An Essay Concerning Human Understanding* (London: T. Tegg and Son, 1836), 36.

14. John F. Walzer, "A Period of Ambivalence: Eighteenth-Century American Childhood," in Lloyd deMause, ed., *The History of Childhood* (Lanham, MD: Rowan & Littlefield Publishers, 1974), 358.

15. Diary entries, May 18, May 21, May 22, 1783, Ethel Armes, ed., *Nancy Shippen: Her Journal Book* (Philadelphia: J. B. Lippincott, 1935), 146–47.

16. Linda Baumgarten, *Eighteenth-Century Clothing at Williamsburg* (Williamsburg, VA: Colonial Williamsburg Foundation, 1986), 72–74.

17. Baumgarten, 43.

18. Baumgarten, 15–16.

19. Turner, *The Mother of Washington*, 115.

20. Eubank, *The Authentic Guide Book of Historic Northern Neck of Virginia*, 93–94.

21. Conway, *Barons of the Potomack and the Rappahannock*, 56–57.

22. Freeman, *George Washington*, vol. 1, 49.

23. Smith, *Inside the Great House*, 45.

24. Janet Golden, *A Social History of Wet Nursing in America: From Breast to Bottle* (New York: Cambridge University Press, 1996), 11–13, 19.

25. Lengel, *First Entrepreneur*, 12.

26. Golden, *A Social History of Wet Nursing in America*, 25.

27. *Virginia Gazette*, December 23, 1775 and April 18, 1777, Colonial Williamsburg Archives.

28. Golden, *A Social History of Wet Nursing in America*, 25.

29. Thompson, *"In the Hands of a Good Providence,"* 20.

30. Custis, *Recollections and Private Memoirs of Washington*, 141.

31. George Washington, *"Letter to the Hebrew Congregations of Newport,"* 1790.

32. John C. Fitzpatrick, ed., *The Writings of George Washington*, vol. 27 (Washington, DC: United States Printing Office, 1938), 1.

33. Thompson, *"In the Hands of a Good Providence,"* 21.

34. Thompson, *"In the Hands of a Good Providence,"* 192; Edward Charles McGuire,

The Religious Opinion and Character of George Washington (New York: Harper & Brothers, 1836), 47–48.

35. Irving, *Life of George Washington*, vol. 1, 22.

36. Thompson, *"In the Hands of a Good Providence,"* 21–22.

37. Robert Shackleton, *The Book of Washington* (Philadelphia: Penn Publishing, 1923), 332.

38. William Jones Rhees, *Visitor's Guide to the Smithsonian Institution and United States National Museum in Washington* (Washington, DC: Judd & Detweiler, 1885), 26.

39. Helen M. Richardson, "George Washington's Christening Robe," *The Churchman*, February 20, 1904, 24.

40. "Washington's Baptism," *Time*, September 5, 1932.

41. Washington family Bible; Mount Vernon Archives, Mount Vernon, Virginia.

42. Letter from Hannah Fairfax Washington to George Washington, April 9, 1792, *PGWPres*, ed. Robert F. Haggard and Mark A. Mastromarino, vol. 10 (Charlottesville: University of Virginia Press, 2002), 240–42.

43. Philip Levy, *Where the Cherry Tree Grew: The Story of Ferry Farm, George Washington's Boyhood Home* (New York: St. Martin's Press, 2013), 39.

44. Luke Pecoraro, in discussion with Scott Mauer, March 15, 2018.

45. Augustine Washington, "Articles of Agreement and Copartnership," April 15, 1737, George Washington papers, Manuscripts and Archives Division, the New York Public Library, New York City, New York.

46. Conway, *Barons of the Potomack and the Rappahannock*, 57–64.

47. Truro Parish Colonial Vestry Book, 47, Pohick Episcopal Church Archives, Lorton, Virginia.

48. John Brooke, *King George III* (London: Constable, 1972), 1.

49. *Virginia Gazette*, June 9, 1738, Colonial Williamsburg Archives.

50. Turner, *The Mother of Washington*, 124–25.

51. Letter from Buckner Stith to George Washington, March 22, 1787, *PGWCon*, ed. W. W. Abbot, vol. 5 (Charlottesville: University of Virginia Press, 1997), 99–101.

52. Chernow, *Washington: A Life*, 7.

53. Kate Douglas Wiggin and Nora A. Smith, *The Story Hour: A Book for the Home and Kindergarten* (Boston: Houghton Mifflin, 1891), 116–17.

54. Randall, *George Washington: A Life*, 18.

55. Turner, *The Mother of Washington*, 122.

56. *Virginia Gazette*, July 22, 1737, Colonial Williamsburg Archives.

57. Smith, *Inside the Great House*, 49–51.

CHAPTER 6: FREDERICKSBURG

1. *Virginia Gazette*, April 21, 1738, Colonial Williamsburg Archives.

2. Conway, *Barons of the Potomack and the Rappahannock*, 71.

3. George H. S. King, "Washington's Boyhood Home," *William and Mary Quarterly*, vol. 16, no. 2, April 1937, 269–73.

4. Freeman, *George Washington*, vol. 1, 58.

5. Turner, *The Mother of Washington*, 128.

6. Turner, 129.

7. Diary entry, January 16, 1760, *DGW*, ed. Donald Jackson, vol. 1 (Charlottesville: University of Virginia Press, 1976), 224–25.

8. *Virginia Gazette*, November 5, 1772, Colonial Williamsburg Archives.

9. Silvanus Jackson Quinn, *The History of the City of Fredericksburg, Virginia* (Richmond, Hermitage Press, 1908), 43.

10. Freeman, *George Washington*, vol. 1, 58; King, "Washington's Boyhood Home," 268.

11. Paul Wilstach, *Mount Vernon: Washington's Home and the Nation's Shrine* (Garden City, New York: Doubleday, Page, 1916), 14.

12. *Virginia Gazette*, August 28, 1746, Colonial Williamsburg Archives.

13. Felder, *Fielding Lewis and the Washington Family*, 83–84; William A. Kretzschmar Jr., et al., eds., *Handbook of the Linguistic Atlas of the Middle and South Atlantic States* (Chicago: University of Chicago Press, 1994), 310.

14. Philip Levy, *Where the Cherry Tree Grew*, 42.

15. Weems, *The Life of George Washington*, 14.

16. Levy, *Where the Cherry Tree Grew*, 214–17.

17. Weems, *The Life of George Washington*, 16–17.

18. Turner, *The Mother of Washington*, 131.

19. Smith, *Inside the Great House*, 82–83.

20. Smith, 103.

21. Smith, 84.

22. Letter from Thomas Jefferson to Walter Jones, January 2, 1814, *PTJRet*, ed. J. Jefferson Looney, vol. 7 (Princeton: Princeton University Press, 2010), 100–104.

23. Pryor, *The Mother of Washington and Her Times*, 148–49; *Virginia Gazette*, August 11, 1774, Colonial Williamsburg Archives.

24. *Virginia Gazette*, May 16, 1745, Colonial Williamsburg Archives.

25. Letter from Robert Douglas to George Washington, May 25, 1795, *PGWPres*, ed. Carol S. Ebel, vol. 18 (Charlottesville: University of Virginia Press, 2015), 173–75.

26. Levy, *Where the Cherry Tree Grew*, 191.

27. Levy, 204–5.

28. Levy, 83.

29. Harland, *The Story of Mary Washington*, 56–57.

30. Smith, *Inside the Great House*, 249–50.

31. Smith, 262.

32. Lossing, *Mary and Martha: The Mother and the Wife of George Washington*, 31.

33. Levy, *Where the Cherry Tree Grew*, 55.

34. Sparks, *The Life of George Washington*, 5.

35. Toner, *Wills of the American Ancestors of General George Washington*, 13–16.

36. King, "Washington's Boyhood Home," 269–73.

37. Laura J. Galke, "The Mother of the Father of Our Country: Mary Ball Washington's Genteel Domestic Habits," *Northeast Historical Archaeology*, vol. 38, 2009, 31; Thompson, *"In the Hands of a Good Providence,"* 16.

38. Brown, "General Washington," 5.

39. Pryor, *The Mother of Washington and Her Times*, 51.

40. Washington, "The Mother and Birthplace of Washington," 833.

41. Michelle Hamilton, in discussion with Scott Mauer, March 2, 2018.

42. Lois Green Carr and Lorena S. Walsh, "The Planter's Wife: The Experience of White Women in Seventeenth-Century Maryland," *William and Mary Quarterly*, vol. 34, no. 4, October 1977, 560.

CHAPTER 7: MATRIARCH

1. Pryor, *The Mother of Washington and Her Times*, 93.

2. Richard Norton Smith, *Patriarch: George Washington and the New American Nation* (Boston: Houghton Mifflin, 1993), xx.

3. Custis, *Recollections and Private Memoirs of Washington*, 129–30.

4. Ford, *The True George Washington*, 17.

5. Letter from George Washington to the Citizens of Fredericksburg, February 14, 1784, *PGWCon*, vol. 1, 122–23.

6. Custis, *Recollections and Private Memoirs of Washington*, 131.

7. Smith, *Inside the Great House*, 241.

8. Freeman, *George Washington*, vol. 1, 57.

9. Turner, *The Mother of Washington*, 137–38.

10. "Washington Guided by Jesuit Rules," *American Catholic Historical Researches*, vol. 21, no. 4, October 1904, 151–53.

11. Meade, *Old Churches, Ministers and Families of Virginia*, vol. 2, 89.

12. Humphreys, *Life of General Washington*, 6.

13. George Washington's school copybook, George Washington Papers, Series 1, Exercise Books, Diaries, and Surveys, Library of Congress, Washington, DC.

14. Lillback and Newcombe, *George Washington's Sacred Fire*, 115.

15. Sparks, *The Life of George Washington*, 6.

16. Angelo, *First Mothers*, 404.

17. "Washington's Copy of Rules of Civility & Decent Behavior in Company and Conversation—Transcription," *The Papers of George Washington*, University of Virginia, http://gwpapers.virginia.edu/documents_gw/civility/civility_transcript .html.

18. Carson Holloway, "It's Time to Rediscover George Washington's Greatness," *Daily Signal*, February 16, 2015.

19. John Frederick Schroeder, ed., *Maxims of Washington; Political, Social, Moral, and Religious* (New York: D. Appleton, 1859).

20. Custis, *Recollections and Private Memoirs of Washington*, 131–34; Pryor, *The Mother of Washington and Her Times*, 36–38.

21. *Virginia Gazette*, January 18, 1740, Colonial Williamsburg Archives.

22. Freeman, *George Washington*, vol. 1, 66–71.

23. Peter R. Henriques, "Major Lawrence Washington versus the Reverend Charles Green: A Case Study of the Squire and the Parson," *Virginia Magazine of History and Biography*, vol. 100, no. 2, April 1992, 233–64.

24. Moore, *The Family Life of George Washington*, 5.

25. Conway, *Barons of the Potomack and the Rappahannock*, 238.

26. Humphreys, *Life of General Washington*, 8; Irving, *Life of George Washington*, vol. 1, 27.

27. Conway, *Barons of the Potomack and the Rappahannock*, 235–40.

28. Harland, *The Story of Mary Washington*, 79–80.

29. Arthur N. Gilbert, "Buggery and the British Navy, 1700–1861," *Journal of Social History*, vol. 10, no. 1, Fall 1976, 74.

30. Sparks, *The Life of George Washington*, 15.

31. Custis, *Recollections and Private Memoirs of Washington*, 131.

32. Chernow, *Washington: A Life*, 18.

33. Custis, *Recollections and Private Memoirs of Washington*, 130.

34. Lossing, *Mary and Martha: The Mother and the Wife of George Washington*, 41.

35. William Quentin Maxwell, "A True State of the Smallpox in Williamsburg, February 22, 1748," *Virginia Magazine of History and Biography*, vol. 63, no. 3, July 1955, 269–74.

36. Diary entry, November 17, 1751, *DGW*, ed. Donald Jackson, vol. 1 (Charlottesville: University of Virginia Press, 1976), 82.

37. Diary entry, November 4, 1751, *DGW*, vol. 1, 72–73.

38. *Virginia Gazette*, September 19, 1777, Colonial Williamsburg Archives.

39. As quoted in John Corry, *The Life of George Washington* (New York: McCarty & White, 1809), 7.

40. Lengel, *First Entrepreneur*, 17.

41. Lengel, 19.

42. Diary entries, March 15 and March 16, 1748, *DGW*, vol. 1, 6–16.

43. As quoted in Mitchell, *The Youth of Washington*, 76–77.

44. Cash account book, July 15, 1748; John C. Fitzpatrick, ed., *George Washington, Colonial Traveller, 1732–1775* (Indianapolis: Bobbs-Merrill, 1927), 17; Freeman, *George Washington*, vol. 1, 228.

45. Deed for Ferry Farm land, July 7, 1748, *PGWCol*, ed. W. W. Abbot, vol. 1 (Charlottesville: University of Virginia Press, 1983), 5; letter from George Washington to Lawrence Washington, May 5, 1749, *PGWCol*, vol. 1, 6–8.

46. For example, she lent 3 pounds, 9 shillings to his music master on September 10, 1748; cash account book, July 15, 1748; Fitzpatrick, ed., *George Washington, Colonial Traveller*, 18.

47. Bill from Dr. John Sutherland to Mary Washington, April 27, 1752, W-1310/A, Mount Vernon Ladies' Association, Mount Vernon, Virginia.

48. Account with Mary Washington and James Buchannen, May–July 1765, W-1310/A, Mount Vernon Ladies' Association, Mount Vernon, Virginia.

49. Letter from George Mason to George Washington, July 29, 1752, *PGWCol*, vol. 1, 52–53.

50. Letter from George Washington to Lawrence Washington, May 5, 1749, *PGWCol*, vol. 1, 6–8.

51. Toner, *Wills of the American Ancestors of General George Washington*, 16–19.

CHAPTER 8: LIEUTENANT COLONEL WASHINGTON

1. Randall, *George Washington: A Life*, 68–69.
2. Carmichael, *Mary Ball Washington*, 25.
3. Letter from Daniel Campbell to George Washington, June 28, 1754, *PGWCol*, vol. 1, 151–53.
4. Letter from William Fairfax to George Washington, June 28, 1755, *PGWCol*, vol. 1, 319.
5. Letter from George Washington to John Augustine Washington, May 31, 1754, *PGWCol*, vol. 1, 118–19.
6. Turner, *The Mother of Washington*, 171.
7. Chernow, *Washington: A Life*, 53.
8. Pryor, *The Mother of Washington and Her Times*, 118.
9. Harland, *The Story of Mary Washington*, 87–88.
10. Letter from George Washington to Robert Orme, April 2, 1755, *PGWCol*, vol. 1, 246–48; Chernow, *Washington: A Life*, 53–54.
11. Irving, *Life of George Washington*, vol. 1, 142.
12. Brockett and Rock, *A Concise History of the City of Alexandria, Va.*, 14.
13. Letter from George Washington to Mary Ball Washington, May 6, 1755, *PGWCol*, vol. 1, 268–69.
14. Letter from George Washington to Mary Ball Washington, June 7, 1755, *PGWCol*, vol. 1, 304–5.
15. Humphreys, *Life of General Washington*, 18.
16. Letter from George Washington to John Augustine Washington, July 18, 1755, *PGWCol*, vol. 1, 343.
17. Letter from George Washington to Mary Ball Washington, July 18, 1755, *PGWCol*, vol. 1, 336–38.
18. Letter from George Washington to Robert Orme, July 28, 1755, *PGWCol*, vol. 1, 347–48.
19. Letter from Joseph Ball to George Washington, September 5, 1755, *PGWCol*, vol. 2, 15–16.
20. Freeman, *George Washington*, vol. 2 (New York: Charles Scribner's Sons, 1948), 114.
21. Letter from George Washington to Mary Ball Washington, August 14, 1755, *PGWCol*, vol. 1, 359–60.
22. Irving, *Life of George Washington*, vol. 1, 193.
23. Letter from George Washington to Mary Ball Washington, September 30, 1757, *PGWCol*, vol. 4, 430–31.
24. General ledger A, cash account, May 17, 1757, Library of Congress, Washington, DC.
25. William Johnson, *George Washington the Christian* (New York: Abingdon Press, 1919), 42–46.
26. Letter from Mary Ball Washington to Joseph Ball, July 26, 1759; as quoted in Hayden, *Virginia Genealogies*, 81.
27. James Thomas Flexner, *George Washington and the New Nation, 1783–1793* (Boston: Little, Brown, 1970), 228.
28. Pryor, *The Mother of Washington and Her Times*, 171–72.

29. Freeman, *George Washington*, vol. 1, 103.

30. Freeman, *George Washington*, vol. 1, 103; Robert Beverley, *The History and Present State of Virginia* (Richmond: J. W. Randolph, 1855), 237, 242.

31. Freeman, *George Washington*, vol. 1, 103; King, "Washington's Boyhood Home," 270.

32. Harland, *The Story of Mary Washington*, 63.

33. Michelle Hamilton, in discussion with Scott Mauer, April 10, 2018.

34. Freeman, *George Washington*, vol. 1, 193.

35. Michelle Hamilton, in discussion with Scott Mauer, April 24, 2018.

36. Gary Hayes, in discussion with Scott Mauer, April 25, 2018.

37. Washington, "The Mother and Birthplace of Washington," 834.

38. Michelle Hamilton, *Mary Ball Washington: The Mother of George Washington* (Ruther Glen, VA: MLH Publications, 2017), 15.

39. *Virginia Gazette*, June 15, 1769, July 20, 1769, Colonial Williamsburg Archives.

40. Court case, December 3, 1751, *PGWCol*, vol. 1, 48–49.

41. Galke, "The Mother of the Father of Our Country," 36.

42. Pryor, *The Mother and Washington and Her Times*, 169; William H. Snowden, *Some Old Historic Landmarks of Virginia and Maryland* (Alexandria, Virginia: G. H. Ramey & Son, 1904), 121.

43. Galke, "The Mother of the Father of Our Country," 39–40; Levy, *Where the Cherry Tree Grew*, 207.

44. Galke, "The Mother of the Father of Our Country," 37–38.

45. Letter from Joseph Ball to Betty Washington, November 2, 1749; as quoted in Paula S. Felder, *Fielding Lewis and the Washington Family*, 68.

46. Levy, *Where the Cherry Tree Grew*, 70.

47. Ford, *The True George Washington*, 21–22.

48. Felder, *Fielding Lewis and the Washington Family*, 60.

49. Letter from George Washington to John Augustine Washington, January 16, 1783, George Washington Papers, series 2, letterbook 11; manuscript/mixed material, Library of Congress, Washington, DC.

50. Letter from George Washington to Henry Knox, April 27, 1787, *PGWCon*, ed. W. W. Abbot, vol. 5 (Charlottesville: University of Virginia Press, 1997), 157–59.

51. Pryor, *The Mother of Washington and Her Times*, 169.

52. Letter from Mary Ball Washington to John Augustine Washington, no date; Mary Washington House Archives, Fredericksburg, Virginia; Michelle Hamilton, in discussion with Scott Mauer, April 24, 2018.

53. Ford, *The True George Washington*, 27.

54. Michelle Hamilton, in discussion with Scott Mauer, April 24, 2018; Katie King, in discussion with Scott Mauer, April 24, 2018.

55. Pryor, *The Mother of Washington and Her Times*, 186.

56. Diary entries, January 16 and 17, 1760, *DGW*, vol. 1, 224–26.

57. General ledger A, cash account, June 4, 1760, *PGWCol*, vol. 6, 429–31.

58. Diary entries, March 6, 1769, November 1, 1769, *DGW*, vol. 2, 132, 193.

59. *Virginia Gazette*, December 11, 1766, December 24, 1767, Colonial Williamsburg Archives.

60. Carmichael, *Mary Ball Washington*, 26.

61. Turner, *The Mother of Washington*, 206.

62. Michelle Hamilton, in discussion with Scott Mauer, April 24, 2018; The fan and case are now housed at the Mary Washington House in Fredericksburg, Virginia.

63. Diary entries, June 25, June 26, July 31, August 1, August 9, 1770, *DGW*, vol. 2, 249, 257, 260, 262.

64. Freeman, *George Washington*, vol. 3 (New York: Charles Scribner's Sons, 1951), 280.

65. Michelle Hamilton, in discussion with Scott Mauer, April 24, 2018.

66. Diary entry, September 13, 1771, *DGW*, vol. 3, 53.

67. Lawrence Martin, ed., *The George Washington Atlas* (Washington, DC: George Washington Bicentennial Commission, 1932), 9.

68. Hugh Mercer ledger, December 1771, Hugh Mercer Apothecary Shop, Fredericksburg, Virginia.

69. Genevieve Bugay, in discussion with Scott Mauer, April 24, 2018; Hugh Mercer ledger, December 1771; Hugh Mercer Apothecary Shop, Fredericksburg, Virginia.

70. Michelle Hamilton, in discussion with Scott Mauer, April 24, 2018.

71. Michelle Hamilton, in discussion with Scott Mauer, April 24, 2018.

72. Diary entry, January 27, 1772, *DGW*, vol. 3, 84.

73. Freeman, *George Washington*, vol. 3, 281.

74. "Evaluations of Sundries belonging to Mary Washington," October 15, 1771, A-415.1, Mount Vernon Ladies' Association, Mount Vernon, Virginia.

75. Diary entry, April 11, 1772, *DGW*, vol. 3, 102.

76. *Virginia Gazette*, November 5, 1772, Colonial Williamsburg Archives.

77. Katie King and Daniel Hawkins, in discussion with Scott Mauer, April 24, 2018.

78. Gary Hayes, in discussion with Scott Mauer, April 26, 2018.

79. *Virginia Gazette*, August 6, 1772, Colonial Williamsburg Archives.

80. Turner, *The Mother of Washington*, 222.

81. Angelo, *First Mothers*, 404.

82. Letter from Edward Jones to George Washington, December 7, 1772, *PGWCol*, vol. 9, 137–38.

83. Letter from George Washington to Betty Washington Lewis, September 13, 1789, *PGWPres*, ed. Dorothy Twohig, vol. 4 (Charlottesville: University of Virginia Press, 1993), 32–36.

84. Flexner, *George Washington: The Forge of Experience*, 265.

CHAPTER 9: BEFORE THE REVOLUTION

1. Emily J. Salmon and Edward D. C. Campbell Jr., eds., *The Hornbook of Virginia History* (Richmond: Library of Virginia, 1994), 103–6.

2. Corry, *The Life of George Washington*, 23.

3. John Howe, *Language and Political Meaning in Revolutionary America* (Boston: University of Massachusetts Press, 2004), 43.

4. Jay A. Parry and Andrew M. Allison, *The Real George Washington* (American Classic series; Malta, ID: National Center for Constitutional Studies, 2010), 95–97.

5. Brooke, *King George III*, 88, 390.

6. *Virginia Gazette*, September 19, 1766, Colonial Williamsburg Archives.

7. James L. Stokesbury, *A Short History of the American Revolution* (New York: HarperCollins, 1991), 29.

8. As quoted in Brockett and Rock, *A Concise History of the City of Alexandria, Va.*, 16.

9. Stokesbury, *A Short History of the American Revolution*, 30.

10. Letter from George Washington to Lord Botetourt, December 8, 1769, *PGW-Col*, ed. W. W. Abbot and Dorothy Twohig, vol. 8 (Charlottesville: University of Virginia Press, 1993), 272–77.

11. Letter from Richard Jackson to Benjamin Franklin, November 12, 1763, *PBF*, ed. Leonard W. Labaree, vol. 10 (New Haven: Yale University Press, 1959), 368–72; letter from Benjamin Franklin to Richard Jackson, June 25, 1763, *PBF*, vol. 11, 234–24.

12. Kevin Phillips, *1775: A Good Year for Revolution* (New York: Penguin Books, 2012), 116.

13. Brooke, *King George III*, 125.

14. Franklin, *The Autobiography of Benjamin Franklin*, 181.

15. As quoted in Corry, *The Life of George Washington*, 28.

16. Claire Priest, "The Stamp Act and the Political Origins of American Legal and Economic Institutions," *Yale Law School Faculty Scholarship Series*, no. 4934, 2015, 886–87.

17. Letter from Benjamin Franklin to Governor William Shirley, December 4, 1754, *PBF*, vol. 5, 443–44.

18. Letter from George Washington to Robert Cary & Company, September 20, 1765, *PGWCol*, vol. 7, 398–402.

19. Letter from George Washington to Francis Dandridge, September 20, 1765, *PGWCol*, vol. 7, 395–96.

20. Brockett and Rock, *A Concise History of the City of Alexandria, Va.*, 17.

21. Chernow, *Washington: A Life*, 137–38.

22. *Virginia Gazette*, May 16, 1766, Colonial Williamsburg Archives.

23. Letter from Benjamin Franklin to Lord Kames, February 25, 1767, *PBF*, vol. 14, 62–71.

24. Corry, *The Life of George Washington*, 31; Phillips, *1775*, 95.

25. "Resolves of the House of Burgesses, Passed the 16th of May, 1769," Library of Virginia Special Collections, Richmond, Virginia.

26. Virginia Nonimportation Resolutions, 17 May 1769, *PTJ*, ed. Julian P. Boyd, vol. 1 (Princeton: Princeton University Press, 1950), 27–31.

27. Letter from George Washington to George Mason, April 5, 1769, *PGWCol*, vol. 8, 177–81; Parry and Allison, *The Real George Washington*, 101.

28. *Virginia Gazette*, March 29, 1770, Colonial Williamsburg Archives.

29. *Virginia Gazette*, April 5, 1770, Colonial Williamsburg Archives.

30. *Virginia Gazette*, September 20, 1770, Colonial Williamsburg Archives.

31. Carmichael, *Mary Ball Washington*, 27.

32. Diary entry, April 11, 1772, *DGW*, ed. Donald Jackson, vol. 3 (Charlottesville: University of Virginia Press, 1976), 102.

33. Diary entry, September 14–17, 1772, *DGW,* vol. 3, 130–31; General ledger B, cash accounts, September 1772, *PGWCol,* vol. 9, 91–92.

34. Diary entry, November 27, 1772, *DGW,* vol. 3, 144; cash accounts, November 1772, *PGWCol,* vol. 9, 118.

35. Corry, *The Life of George Washington,* 34–35.

36. Corry, 34–35.

37. *Virginia Gazette,* September 1, 1774, Colonial Williamsburg Archives.

38. Diary entry, July 29, 1770, *DGW,* vol. 2, 256.

39. Flexner, *George Washington: The Forge of Experience,* 320.

40. Corry, *The Life of George Washington,* 35.

41. *Virginia Gazette,* May 26, 1774, Colonial Williamsburg Archives.

42. Diary entry, June 1, 1774, *DGW,* vol. 3, 254.

43. Letter from John Harrower to Mrs. Harrower, June 14, 1744; "Diary of John Harrower, 1773–1776," *The American Historical Review,* vol. 6, no. 1, October 1900, 84; Hamilton, *Mary Ball Washington,* 20–21.

44. Receipt from Robert Broom to Mary Washington, May 18, 1774; Mary Washington House Archives, Fredericksburg, Virginia.

45. Turner, *The Mother of Washington,* 211.

46. Letter from George Washington to George William Fairfax, June 10–15, 1774, *PGWCol,* vol. 10, 94–101.

47. Letter from George Washington to Bryan Fairfax, July 20, 1774, *PGWCol,* vol. 10, 128–31.

48. Diary entries, August 23 and 25, 1774, Phillip Vickers Fithian, *Journal and Letters, 1767–1774* (Carlisle, MA: Applewood Books, 1900), 234–35.

49. *Virginia Gazette,* July 21, 1774, Colonial Williamsburg Archives.

50. As quoted in Chernow, *Washington: A Life,* 171.

51. Letter from George Washington to Thomas Jefferson, August 5, 1774, *PGWCol,* vol. 10, 142–43.

52. Diary entry, August 9, 1774, *DGW,* vol. 3, 269; cash accounts, August 1774, *PGWCol,* vol. 10, 138–41.

53. Wilson, *George Washington,* 154.

54. As quoted in Douglas Bradburn, *The Citizenship Revolution: Politics & the Creation of the American Union, 1774–1804* (Charlottesville: University of Virginia Press, 2009), 19.

55. Bradburn, *The Citizenship Revolution,* 23; diary entry, September 8, 1774, *Diary and Autobiography of John Adams,* ed. L. H. Butterfield, vol. 2 (Cambridge, Massachusetts: Harvard University Press, 1961), 128–31.

56. As quoted in Bradburn, *The Citizenship Revolution,* 20.

57. Worthington Chauncey Ford, ed., *Journals of the Continental Congress, 1774–1789* (Washington, DC: Government Printing Office, 1905), vol. 1, 67–68.

58. Chernow, *Washington: A Life,* 173.

59. Letter to Robert McKenzie, October 9, 1774; as quoted in Chernow, *Washington: A Life,* 174.

60. General ledger B, card playing expenses, 1772–1774, *PGWCol,* vol. 10, 223.

61. General ledger B, cash accounts, October 1774, *PGWCol,* vol. 10, 166–68.

62. Diary entry, October 11, 1774, Fithian, *Journal and Letters*, 266.

63. *Virginia Gazette*, August 11, 1774, Colonial Williamsburg Archives.

64. *Virginia Gazette*, October 20, 1774, Colonial Williamsburg Archives.

65. *Virginia Gazette*, August 11, 1774, August 18, 1774, Colonial Williamsburg Archives.

66. *Virginia Gazette*, November 3, 1774, Colonial Williamsburg Archives.

67. *Virginia Gazette*, December 1, 1774, Colonial Williamsburg Archives.

68. Account of the weather in December 1774, January 1775, and February 1775, *DGW*, vol. 3, 300–30, 305–7, 310–11.

69. Diary entry, March 23, 1775, *DGW*, vol. 3, 318.

70. As quoted in Freeman, *George Washington*, vol. 3, 403–4.

71. Freeman, *George Washington*, vol. 3, 406–7.

72. Account with Mary Washington, April 27, 1775, *PGWCol*, vol. 10, 347–49.

73. Letter from George Washington to Bryan Fairfax, March 1, 1778, *PGWRev*, vol. 14, ed. David R. Hoth (Charlottesville: University of Virginia Press, 2004), 9–11.

74. Letter from George Washington to Joseph Jones, February 11, 1783; manuscript/mixed material, Library of Congress, Washington, DC.

CHAPTER 10: OFF TO WAR

1. *Virginia Gazette*, May 5, 1775, Colonial Williamsburg Archives.

2. Freeman, *George Washington*, vol. 3, 412–13; Chernow, *Washington: A Life*, 181.

3. Letter from George Washington to George William Fairfax, May 31, 1775, *PGWCol*, vol. 10, 367–68; Chernow, *Washington: A Life*, 181.

4. Freeman, *George Washington*, vol. 3, 419.

5. Diary entry, in Congress, June and July 1775, *Diary and Autobiography of John Adams*, vol. 3, 321.

6. Diary entry, *Diary and Autobiography of John Adams*, 321.

7. Diary entry, in Congress, June and July 1775, *Diary and Autobiography of John Adams*, vol. 3, 321–24; Ford, *Journals of the Continental Congress*, 91.

8. Diary entry, June 15, 1775, *DGW*, vol. 3, ed. Donald Jackson (Charlottesville: University of Virginia Press, 1978), 336.

9. Ford, *Journals of the Continental Congress*, vol. 2, 92.

10. Freeman, *George Washington*, vol. 3, 439–40.

11. Letter from George Washington to Martha Washington, June 18, 1775, *PGWRev*, vol. 1, ed. Philander D. Chase (Charlottesville: University of Virginia Press, 1985), 3–6.

12. Letter from George Washington to Burwell Bassett, June 19, 1775, *PGWRev*, vol. 1, 12–14.

13. Letter from George Washington to John Parke Custis, June 19, 1775, *PGWRev*, vol. 1, 15–16.

14. Letter from George Washington to John Augustine Washington, June 20, 1775, *PGWRev*, vol. 1, 19–20.

15. Letter from George Washington to Martha Washington, June 20, 1775, *PGWRev*, vol. 1, 27–28.

16. *Virginia Gazette*, July 6, 1775, Colonial Williamsburg Archives.

17. *Virginia Gazette*, May 13, 1775, Colonial Williamsburg Archives.

18. Custis, *Recollections and Private Memoirs of Washington*, 135.

19. Harland, *The Story of Mary Washington*, 108.

20. Conkling, *Memoirs of the Mother and Wife of Washington*, 43.

21. Michelle Hamilton, in discussion with Scott Mauer, April 24, 2018.

22. Letter from George Washington to Samuel Washington, July 20, 2018, *PGWRev*, vol. 1, 134–36.

23. Ford, *Journals of the Continental Congress*, vol. 2, 158–62.

24. "His Majesty's Most Gracious Speech to Both Houses of Parliament," October 27, 1775; Rare Book and Special Collections Division, Library of Congress, Washington, DC.

25. John E. Selby, *The Revolution in Virginia, 1775–1783* (Williamsburg: Colonial Williamsburg Foundation, 1988), 62–65.

26. *Virginia Gazette*, November 25, 1775, Colonial Williamsburg Archives.

27. *Virginia Gazette*, November 25, 1775, Colonial Williamsburg Archives.

28. Selby, *The Revolution in Virginia*, 67.

29. Stokesbury, *A Short History of the American Revolution*, 87.

30. *Virginia Gazette*, September 27, 1776, Colonial Williamsburg Archives.

31. *Virginia Gazette*, February 2, 1776, Colonial Williamsburg Archives.

32. *Virginia Gazette*, July 12, 1776, Colonial Williamsburg Archives.

33. Ford, *Journals of the Continental Congress*, vol. 8, 464.

34. John Mollo, *Uniforms of the American Revolution, 1775–1781* (New York: Sterling, 1975), 51–59; Stokesbury, *A Short History of the American Revolution*, 46.

35. Stokesbury, *A Short History of the American Revolution*, 69–70.

36. Clary, *George Washington's First War*, 127–28.

37. Mollo, *Uniforms of the American Revolution*, 11–13.

38. Stokesbury, *A Short History of the American Revolution*, 72.

39. Samuel Seabury, *A View of the Controversy Between Great-Britain and Her Colonies* (New York: Royal Exchange, 1775), 13–14.

40. Mollo, *Uniforms of the American Revolution*, 33.

41. Letter from George Washington to William Shippen Jr., February 6, 1777, *PGWRev*, vol. 8, 264.

42. Volo and Volo, *Daily Life During the American Revolution*, 230.

43. *Virginia Gazette*, January 13, 1776, Colonial Williamsburg Archives.

44. *Virginia Gazette*, September 15, 1774, Colonial Williamsburg Archives.

45. Petula Dvorak, "This Woman's Name Appears on the Declaration of Independence. So Why Don't We Know Her Story?" *Washington Post*, July 3, 2017.

46. *Virginia Gazette*, September 21, 1776, Colonial Williamsburg Archives.

47. Volo and Volo, *Daily Life During the American Revolution*, 232.

48. Volo and Volo, 233.

49. *Virginia Gazette*, April 18, 1777, Colonial Williamsburg Archives.

50. *Virginia Gazette*, July 3, 1779, Colonial Williamsburg Archives.

51. *Virginia Gazette*, May 15, 1779, Colonial Williamsburg Archives.

52. Volo and Volo, *Daily Life During the American Revolution*, 240.

53. Selby, *The Revolution in Virginia*, 235–36.

54. "Sketch of Washington," *New York Times*, June 27, 1856.

55. Custis, *Recollections and Private Memoirs of Washington*, 413–14.

56. As quoted in Volo and Volo, *Daily Life During the American Revolution*, 243.

57. Custis, *Recollections and Private Memoirs of Washington*, 139–40.

58. Pryor, *The Mother of Washington and Her Times*, 256.

59. Diary entry, July 14, 1782, Howard C. Rice and Anne S. K. Brown, eds., *The American Campaigns of Rochambeau's Army, 1780–1783*, vol. 1 (Princeton: Princeton University Press, 1972), 73.

60. As quoted in Chernow, *Washington: A Life*, 396.

61. Michelle Hamilton, in discussion with Scott Mauer, April 24, 2018.

62. Freeman, *George Washington*, vol. 5 (New York: Charles Scribner's Sons, 1952), 491.

63. Harland, *The Story of Mary Washington*, 112.

64. Custis, *Recollections and Private Memoirs of Washington*, 137–38.

65. Michelle Hamilton, in discussion with Scott Mauer, April 24, 2018.

66. Sparks, *The Life of George Washington*, 284–85.

67. Lossing, *Mary and Martha: The Mother and the Wife of Washington*, 58.

68. *Virginia Gazette*, January 17 and 31, 1777, Colonial Williamsburg Archives.

69. Genevieve Bugay, in discussion with Scott Mauer, April 24, 2018.

70. As quoted in Rupert Hughes, *George Washington: The Savior of the States, 1771–1781* (New York: William Morrow, 1930), 82.

71. Washington, "The Mother and Birthplace of Washington," 837.

72. Thompson, *"In the Hands of a Good Providence,"* 92.

73. Letter from George Washington to Landon Carter, April 15, 1777, *PGWRev*, vol. 9, 170–71.

74. Sparks, *The Life of George Washington*, 329–30.

75. William Whitney Cone, *Some Account of the Cone Family in America* (Topeka, KS: Crane & Company, 1903), 323.

76. Humphreys, *Life of General Washington*, 33.

77. Conkling, *Memoirs of the Mother and Wife of Washington*, 47; Custis, *Recollections and Private Memoirs of Washington*, 137.

78. Hughes, *George Washington: The Savior of the States*, 44.

79. Letter from George Weedon to George Washington, March 30, 1778, *PGWRev*, vol. 14, 361–62.

80. Letters from Mary Washington to Lund Washington, December 4 and 23, 1778; Mary Washington House Archives, Fredericksburg, Virginia.

81. Washington, "The Mother and Birthplace of Washington," 838.

82. Letter from Benjamin Harrison to George Washington, February 25, 1781; manuscript/mixed material, Library of Congress, Washington, DC.

83. Letter from George Washington to Benjamin Harrison, March 21, 1781; as quoted in John C. Fitzpatrick, ed., *The Writings of George Washington*, vol. 21 (Washington, DC: United States Printing Office, 1937), 340–42.

84. Letter from Marquis de Lafayette to George Washington, April 8, 1781;

Memoirs, Correspondence and Manuscripts of General Lafayette, vol. 1 (London: Saunders and Otley, 1837), 397.

85. Letter from Marquis de Lafayette to George Washington, February 5, 1783, *Memoirs, Correspondence and Manuscripts of General Lafayette*, vol. 2, 56.

86. Conkling, *Memoirs of the Mother and Wife of Washington*, 47.

87. Turner, *The Mother of Washington*, 241.

88. Chernow, *Washington: A Life*, 422.

89. James Thomas Flexner, *George Washington: In the American Revolution, 1775–1783* (Boston: Little, Brown, 1968), 471.

90. Freeman, *George Washington*, vol. 5, 409.

91. Emily White Fleming, *Historic Periods of Fredericksburg, 1608–1861* (Richmond: W. C. Hill Printing, 1921), 21.

92. Freeman, *George Washington*, vol. 5, 409.

93. Bill from Captain Marban to Mary Washington, September 8, 1783, W-1310/A.27, Mount Vernon Ladies' Association, Mount Vernon, Virginia.

94. Pryor, *The Mother of Washington and Her Times*, 285.

95. Turner, *The Mother of Washington*, 238.

96. Stokesbury, *A Short History of the American Revolution*, 279.

97. Hughes, *George Washington: The Savior of the States*, 119.

98. Hill, *Ball Families of Virginia's Northern Neck*, 20.

99. Benjamin Tallmadge, *Memoir of Col. Benjamin Tallmadge* (New York: Thomas Holman, Book and Job Printer, 1858), 63–64.

CHAPTER 11: A SEPARATE PEACE

1. Humphreys, *Life of General Washington*, 37.

2. Flexner, *George Washington and the New Nation, 1783–1793*, 39.

3. Letter from George Washington to Arthur Young, December 12, 1793, *PGWPres*, ed. David R. Hoth, vol. 14 (Charlottesville: University of Virginia Press, 2008), 504–14.

4. Lund Washington, list of escaped slaves, 1781, Mount Vernon Ladies' Association, Mount Vernon, Virginia.

5. Letter from George Washington to Lund Washington, April 30, 1781, manuscript/mixed material, Library of Congress, Washington, DC.

6. Letter from George Washington to Henry Knox, January 5, 1785, *PGWCon*, ed. W. W. Abbot, vol. 2 (Charlottesville: University of Virginia Press, 1992), 253–56.

7. Letter from George Washington to Fielding Lewis Jr., February 27, 1784, *PGWCon*, vol. 1, 161–62.

8. Letter from George Washington to John Augustine Washington, January 16, 1783, manuscript/mixed material, Library of Congress, Washington, DC.

9. Flora Fraser, *George & Martha Washington: A Revolutionary Marriage* (London: Bloomsbury, 2015), 284–86.

10. Custis, *Recollections and Private Memoirs of Washington*, 141.

11. Fleming, *Historic Periods of Fredericksburg*, 22.

12. David M. Matteson, "The Fredericksburg Peace Ball," *Virginia Magazine of History and Biography*, vol. 49, no. 2 (April 1941), 153.

13. Letter from George Washington to Jacob Read, February 12, 1784, *PGWCon*, vol. 1, 112–13.

14. Pryor, *The Mother of Washington and Her Times*, 306.

15. Chernow, *Washington: A Life*, 423.

16. Custis, *Recollections and Private Memoirs of Washington*, 142–43.

17. Letter from Nellie Parke Custis Lewis, March 16, 1851, Mary Washington House Archives, Fredericksburg, Virginia.

18. Custis, *Recollections and Private Memoirs of Washington*, 142–43.

19. Harland, *The Story of Mary Washington*, 123–24.

20. Harland, 123.

21. Letter from Citizens of Fredericksburg to George Washington, February 14, 1784, *PGWCon*, vol. 1, 120–22.

22. Lossing, *Mary and Martha: The Mother and the Wife of George Washington*, 64–65.

23. *Mount Vernon*, "Lafayette Gingerbread," http://www.mountvernon.org/inn /recipes/article/lafayette-gingerbread/.

24. Harland, *The Story of Mary Washington*, 128.

25. See, for example, letter from the Marquis de Lafayette to George Washington, February 6, 1786, *PGWCon*, vol. 3, 547; letter from the Marquis de Lafayette to George Washington, October 15, 1787, *PGWCon*, vol. 5, 378; and letter from the Marquis de Lafayette to George Washington, May 25, 1788, *PGWCon*, vol. 6, 295.

26. Ledger B, February 15, 1784; Washington Papers, University of Virginia, Charlottesville, Virginia.

27. *Virginia Gazette, or The American Advertiser,* February 21, 1784, Newspaper Vault VA Vault, Library of Congress Archives, Washington, DC.

28. Letter from George Washington to the Citizens of Fredericksburg, February 14, 1784, *PGWCon*, vol. 1, 122–23.

29. Chernow, *Washington: A Life*, 463.

30. Letter from George Washington to Daniel Boinod and Alexander Gaillard, February 18, 1784, *PGWCon*, vol. 1, 126–27.

31. Letter from George Washington to Elias Boundinot, February 18, 1784, *PGWCon*, vol. 1, 127–28.

32. Letter from George Washington to Annis Boundinot, February 18, 1784, *PGWCon*, vol. 1, 132–33.

33. Letter from William Simmes, June 23, 1783; Mary Washington House Archives, Fredericksburg, Virginia.

34. Diary entry, April 29, 1785, *DGW*, ed. Donald Jackson and Dorothy Twohig, vol. 4 (Charlottesville: University of Virginia Press, 1978), 131.

35. Resolutions of the Dismal Swamp Company, May 2, 1785, *PGWCon*, vol. 2, 530–31.

36. Diary entry, May 5, 1785, *DGW*, vol. 4, 134.

37. Diary entry, February 18, 1786, *DGW*, vol. 4, 276–83; Washington's slave list, June 1799, *PGWRet*, ed. W. W. Abbot, vol. 4 (Charlottesville: University of Virginia Press, 1999), 527–42.

38. Letter from George Washington to Edward Newenham, April 20, 1787, *PGWCon*, vol. 5, 151–53; Flexner, *George Washington and the New Nation*, 43.

39. Diary entry, April 24, 1786, *DGW,* vol. 4, 316–17.

40. Letter from Adrienne, Marquise de Lafayette, to George Washington, April 15, 1785, *PGWCon,* vol. 2, 502–3.

41. Letter from George Washington to Adrienne, Marquise de Lafayette, May 10, 1786, *PGWCon,* vol. 4, 39–40.

42. William Waller Hening, ed., *The Statutes at Large,* vol. 12 (Richmond: George Cochran, 1823), 375–76.

43. An Act for Establishing Religious Freedom, 1786; Special Collections, Library of Virginia, Richmond, Virginia.

44. James Byrd, "Was the American Revolution a Holy War?" *Washington Post,* July 7, 2013.

45. Mary Thompson, in discussion with Scott Mauer, May 16, 2018.

CHAPTER 12: HONORED MADAM

1. Letter from Mary Washington to John Augustine Washington, no date; Mary Washington House Archives, Fredericksburg, Virginia.

2. Daniel Blake Smith, "Mortality and Family in the Colonial Chesapeake," *Journal of Interdisciplinary History,* vol. 8, no. 3, Winter 1978, 416–18.

3. Letter from Mary Ball Washington to John Augustine Washington, no date; Mary Washington House Archives, Fredericksburg, Virginia.

4. Michelle Hamilton, in discussion with Scott Mauer, April 24, 2018.

5. Flexner, *George Washington and the New Nation,* 37.

6. Letter from George Washington to Mary Ball Washington, February 15, 1787, *PGWCon,* ed. W. W. Abbot, vol. 5 (Charlottesville: University of Virginia Press, 1997), 33–37.

7. Letter from Marquis de Lafayette to George Washington, February 7, 1787, *PGWCon,* vol. 5, 15.

8. Chernow, *Washington: A Life,* 521–23.

9. Matthew Hale, *Contemplations, Moral and Divine* (London: D. Brown, J. Walthoe, et al., 1711), 182.

10. Hale, *Contemplations, Moral and Divine,* 110–11.

11. Hale, 143.

12. Hale, 163.

13. Diary entry, April 26, 1787, *DGW,* vol. 5, 143–44; letter from George Washington to Henry Knox, April 27, 1787, *PGWCon,* vol. 5, 157–59.

14. Letter from George Washington to Robert Morris, May 5, 1787, *PGWCon,* vol. 5, 171.

15. Michelle Hamilton, in discussion with Scott Mauer, April 24, 2018.

16. Diary entry, April 27, 1787, *DGW,* vol. 5, 145.

17. As quoted in Chernow, *Washington: A Life,* 526.

18. Diary entry, May 25, 1787, *DGW,* vol. 5, 162.

19. Joseph J. Ellis, *His Excellency: George Washington* (New York: Random House, 2004), 177.

20. Flexner, *George Washington and the New Nation,* 118.

21. Flexner, 116.

22. The Virginia Plan, May 29, 1787, *PJM*, ed. Robert A. Rutland, Charles F. Hobson, William M. E. Rachal, and Frederika J. Teute, vol. 10 (Chicago: University of Chicago Press, 1977), 12–18.

23. Reply to the New Jersey plan, 1787, *PJM*, vol. 10, 55–63.

24. Ron Chernow, *Alexander Hamilton* (New York: Penguin Press, 2004), 232.

25. Letter from George Washington to George Augustine Washington, June 3, 1787, *PGWCon*, vol. 5, 217–19.

26. Letter from George Washington to Alexander Hamilton, July 10, 1787, *PGWCon*, vol. 5, 257.

27. Flexner, *George Washington and the New Nation*, 132–33.

28. Diary entry, September 17, 1787, *DGW*, vol. 5, 185.

29. Diary entry, September 22, 1787, *DGW*, vol. 5, 186–87.

30. William Waller Hening, ed., *The Statutes at Large*, vol. 10 (Richmond: George Cochran, 1822), 439–43.

31. Letter from John Dawson to James Madison, June 12, 1787, *PJM*, vol. 10, 47–48.

32. Letter from William Roberts to George Washington, September 2, 1787, *PGWCon*, vol. 5, 308–9.

33. Herbert Baxter Adams, *The Life and Writings of Jared Sparks*, vol. 2 (Cambridge: Riverside Press, 1893), 29.

34. List of taxable property in Fredericksburg, 1787 and 1788, Fredericksburg Circuit Court Archives, Fredericksburg, Virginia.

35. J. Travis Walker, in discussion with Scott Mauer, May 31, 2018.

36. *The Debates in the Several State Conventions, of the Adoption of the Federal Constitution*, ed. Jonathan Elliot, vol. 3 (Washington DC: Government Printing Office, 1836), 176.

37. *The Debates in the Several State Conventions*, 657–59.

38. Pryor, *The Mother of Washington and Her Times*, 365.

39. Diary entry, June 10, 1788, *DGW*, vol. 5, 339–40.

40. General ledger B, June 17, 1788, University of Virginia, Charlottesville, Virginia.

41. Philip Slaughter, *History of St. George's Parish* (Richmond, VA: J. W. Randolph & English, 1890), 25.

42. Letter from George Washington to Henry Knox, June 17, 1788, *PGWCon*, vol. 6, 332; letter from George Washington to Richard Henderson, June 19, 1788, *PGWCon*, vol. 6, 339; letter from George Washington to James Madison, June 23, 1788, *PGWCon*, vol. 6, 351.

43. *The Debates in the Several State Conventions*, vol. 1, 332–33.

CHAPTER 13: UNTO DUST SHALT THOU RETURN

1. Chernow, *Washington: A Life*, 550–51.

2. *Virginia Herald and Fredericksburg Advertiser*, April 23, 1789, Central Rappahannock Heritage Center, Fredericksburg, Virginia.

3. Letter from George Washington to Marquis de Lafayette, April 28, 1788, *PGWCon*, ed. W. W. Abbot. vol. 6 (Charlottesville: University of Virginia Press, 1997), 242–46.

4. Letter from George Washington to Alexander Hamilton, *PGWPres*, ed. Dorothy Twohig, vol. 1 (Charlottesville: University of Virginia Press, 1987), 31–33.

5. Letter from James Madison to Thomas Jefferson, May 23, 1789, *PJM*, vol. 12, ed. Charles F. Hobson and Robert A. Rutland (Charlottesville: University of Virginia Press, 1979), 182–83; letter from Thomas Jefferson to William Carmichael, August 9, 1789, *PTJ*, vol. 15, ed. Julian P. Boyd (Princeton: Princeton University Press, 1958), 336–38.

6. Letter from George Washington to John Langdon, April 14, 1789, *PGWPres*, vol. 2, 54.

7. Letter from George Washington to Richard Conway, March 6, 1789, *PGWPres*, vol. 1, 368–69.

8. *Virginia Herald and Fredericksburg Advertiser*, March 12, 1789, microfilm, Library of Virginia, Richmond, Virginia.

9. Custis, *Recollections and Private Memoirs of Washington*, 145–46.

10. Washington, "The Mother and Birthplace of Washington," 841.

11. General ledger B, March 11, 1789, Washington Papers, University of Virginia, Charlottesville, Virginia.

12. Letter from Robert Lewis to George Washington, March 18, 1789, *PGWPres*, vol. 1, 404.

13. *Virginia Gazette*, November 17, 1738; Colonial Williamsburg Archives, Williamsburg, Virginia.

14. *Virginia Gazette*, June 16, 1768.

15. Letter from Benjamin Rush to Elisha Hall, July 6, 1789; *Letters of Benjamin Rush*, ed. L. H. Butterfield, vol. 1 (Princeton: Princeton University Press, 1951), 518.

16. James E. Guba and Philander D. Chase, "Anthrax and the President, 1789," *Papers of George Washington Newsletter*, no. 5, Spring 2002, 4–6.

17. Letter from Betty Lewis to George Washington, July 24, 1789, *PGWPres*, vol. 3, 301–2.

18. Letter from Burgess Ball to George Washington, August 25, 1789, *PGWPres*, vol. 3, 536–37.

19. *The Book of Common Prayer, and Administration of the Sacraments and Other Rites and Ceremonies of the Church, According to the Use of the Protestant Episcopal Church in the United States of America* (Charleston: W. P. Young, 1808), 219–220.

20. Letter from Burgess Ball to George Washington, August 25, 1789, *PGWPres*, vol. 3, 536–37.

21. Pryor, *The Mother of Washington and Her Times*, 365–67.

22. Custis, *Recollections and Private Memoirs of Washington*, 146.

23. Flexner, *George Washington and the New Nation*, 227.

24. *Gazette of the United States*, September 9, 1789; as quoted in William Spohn Baker, *Washington after the Revolution* (Philadelphia: J. B. Lippincott, 1898), 133–34.

25. Washington Irving, *Life of George Washington*, vol. 5 (New York: G. P. Putnam and Son, 1869), 25.

26. *Virginia Herald and Fredericksburg Advertiser*, August 27, 1789; microfilm, Library of Virginia, Richmond, Virginia.

27. Diary entry, September 1, 1789, Robert Lewis, *Diary, July 4, 1789 to September 1, 1789,* 68–69; Papers of George Washington, Archives of Mount Vernon, Virginia.

28. As quoted in Washington, "The Mother and Birthplace of Washington," 841.

29. Letter from George Washington to Henry Knox, September 4, 1789, *PGWPres,* vol. 3, 600.

30. Chernow, *Washington: A Life,* 589.

31. *Gazette of the United States,* September 8, 1789; as quoted in Baker, *Washington after the Revolution,* 133.

32. Chernow, *Washington: A Life,* 589.

33. Letter from George Washington to Betty Lewis, September 13, 1789, *PGWPres,* vol. 4, 32–36.

34. Statement of money received and paid on the account of the estate of Mary Washington, Virginiana Room, Central Rappahannock Regional Library, Fredericksburg, Virginia.

35. Statement of money received and paid on the account of the estate of Mary Washington, Virginiana Room, Central Rappahannock Regional Library, Fredericksburg, Virginia.

36. Adams, *The Life and Writings of Jared Sparks,* vol. 2, 28–29.

37. Michelle Hamilton, in discussion with Scott Mauer, April 24, 2018.

38. *Virginia Herald and Fredericksburg Advertiser,* October 15, 1789; microfilm, Library of Virginia, Richmond, Virginia.

39. Memorandum of effects not mentioned in Mrs. Washington's will, date unknown, Central Rappahannock Heritage Center, Fredericksburg, Virginia.

40. Letter from Betty Lewis to George Washington, October 1, 1789, *PGWPres,* vol. 4, 122–23, letter from Burgess Ball to Charles Carter, October 8, 1789, *PGWPres,* vol. 4, 146–47.

41. Letter from Burgess Ball to George Washington, December 26, 1789, *PGWPres,* vol. 4, 448–49.

42. Letter from Charles Carter to George Washington, October 14, 1793, *PGWPres,* vol. 14, 204–5.

43. Letter from George Washington to Charles Carter, May 29, 1794, *PGWPres,* vol. 16, 152–53.

EPILOGUE: A MONUMENTAL LEGACY

1. Letter from George Washington to Burgess Ball, September 22, 1799, *PGWRet,* ed. W. W. Abbot, vol. 4 (Charlottesville: University of Virginia Press, 1999), 318.

2. Letter from Hugh Mercer to George Washington, April 6, 1774, *PGWCol,* vol. 10, 23, letter from George Washington to Hugh Mercer, April 11, 1774, *PGWCol,* vol. 10, 27–28.

3. Letter from George Washington to Lund Washington, December 18, 1778, *PGWRev,* ed. Edward G. Lengel, vol. 18 (Charlottesville: University of Virginia Press, 2008), 459–63.

4. Levy, *Where the Cherry Tree Grew,* 195–96.

5. Rick Rojas, "Finding a Lock of George Washington's Hair, and a Link to American History," *New York Times,* February 18, 2018.

6. Schroeder, *Maxims of Washington*, vi–vii.

7. Pryor, *The Mother of Washington and Her Times*, 353.

8. Custis, *Recollections and Private Memoirs of Washington*, 147–48.

9. *Political Arena*, August 8, 1828; microfilm, University of Mary Washington Library, Fredericksburg, Virginia.

10. Flexner, *George Washington and the New Nation*, 228.

11. Pryor, *The Mother of Washington and Her Times*, 355.

12. Sarah Olson, "Historic Furniture Study: The Ancient Kitchen and Colonial Garden, George Washington Birthday National Monument, Virginia," November 1974, 1; George Washington National Monument, Virginia.

13. *Report of the Chief of Engineers, U.S. Army, 1929*, vol. 1 (Washington, DC: Government Printing Office, 1929), 2118–20; Hudson, *George Washington Birthplace: National Monument*, 27–30.

14. As quoted in Conkling, *Memoirs of the Mother and Wife of Washington*, 74–78.

15. Susan Rivère Hetzel, *The Building of a Monument: A History of the Mary Washington Associations and Their Work* (Lancaster, PA: Wickersham, 1903), 45.

16. Hetzel, *The Building of a Monument*, 127–30.

17. *Lisbon Sun*, May 18, 1894, 6.

18. Hetzel, *The Building of a Monument*, 147–48.

19. Hetzel, 151.

20. Anthony Leviero, "Eisenhower Extols Mary Washington," *New York Times*, May 10, 1954, 17.

21. Irving, *Life of Washington*, vol. 5, 26.

Index

About the Author

CRAIG SHIRLEY is the author of *Reagan Rising, Rendezvous with Destiny, Reagan's Revolution, Last Act,* and the *New York Times* bestseller *December 1941.* He is a regular commentator throughout the media and a contributor to national publications, and was hailed by the *London Telegraph* as "the best of the Reagan biographers." He is the Visiting Reagan Scholar at Eureka College, Reagan's alma mater, and lectures often at the Reagan Library and the Reagan Ranch. He and his wife, Zorine, divide their time between Ben Lomond, a three-hundred-year-old Georgian manor house in Essex County, Virginia, and Trickle Down Point, on the Rappahannock River in Lancaster, Virginia. They are the parents of four children: Matthew, Andrew, Taylor, and Mitchell.